T0210557

Lecture Notes in Computer Science 9026

Commenced Publication in 1973
Founding and Former Series Editors:
Gerhard Goos, Juris Hartmanis, and Jan van Leeuwen

More information about this series at http://www.springer.com/series/7407

Gabriela Ochoa · Francisco Chicano (Eds.)

Evolutionary Computation in Combinatorial Optimization

15th European Conference, EvoCOP 2015
Copenhagen, Denmark, April 8–10, 2015
Proceedings

 Springer

Editors
Gabriela Ochoa
University of Stirling
Stirling
UK

Francisco Chicano
University of Málaga
Málaga
Spain

ISSN 0302-9743   ISSN 1611-3349   (electronic)
Lecture Notes in Computer Science
ISBN 978-3-319-16467-0   ISBN 978-3-319-16468-7   (eBook)
DOI 10.1007/978-3-319-16468-7

Library of Congress Control Number: 2015933497

LNCS Sublibrary: SL1 – Theoretical Computer Science and General Issues

Springer Cham Heidelberg New York Dordrecht London

Cover illustration: Designed by Mauro Castelli, ISEGI, Universidade Nova de Lisboa, Portugal

Printed on acid-free paper

Springer International Publishing AG Switzerland is part of Springer Science+Business Media
(www.springer.com)

Preface

Combinatorial optimization is the discipline of decision making dealing with discrete alternatives. The field is at the interface between discrete mathematics, computing science, operational research, and recently also machine learning, and includes a diversity of algorithms and hybrid methods. Stochastic local search (metaheuristics), evolutionary, and other nature-inspired algorithms are a family of methods able to provide robust, high quality solutions to problems of a realistic size in reasonable time. These methods are also relatively simple to design and implement, and offer high flexibility. Many challenging applications in science, industry, and commerce can be formulated as optimization problems. A growing number of them have been successfully solved using the sort of computational methods mentioned above, which are the main content of these proceedings.

EvoCOP was held for the first time in 2001, as the first workshop specifically devoted to evolutionary computation in combinatorial optimization. In 2004 it became a conference, and since then it runs annually. This volume contains the proceedings of EvoCOP 2015, the 15th European Conference on Evolutionary Computation in Combinatorial Optimization, which was held in Copenhagen, Denmark, during April 8–10, 2015. EvoCOP is one of the four events of Evostar 2015. The other three are EuroGP (18th European Conference on Genetic Programming), EvoMUSART (4th International Conference on Evolutionary and Biologically Inspired Music, Sound, Art, and Design), and EvoApplications (18th European Conference on the Applications of Evolutionary Computation, formerly known as EvoWorkshops).

Previous EvoCOP proceedings were published by Springer in the series Lecture Notes in Computer Science (LNCS Volumes 2037, 2279, 2611, 3004, 3448, 3906, 4446, 4972, 5482, 6022, 6622, 7245, 7832, 8600). The table in the next page reports the statistics for each conference.

This year, 19 out of 46 papers were accepted after our rigorous double-blind process, resulting in a 41.3 % acceptance rate, the tighter since 2010. We would like to thank the quality and timeliness of our PC members work, especially since this year's time frame overlapped with the Christmas break. Decisions considered both the reviewers report and evaluation of the Program Chairs. The number of submissions this year shows an increase and we hope this will be maintained as a future trend. The 19 accepted papers covered methodology, applications, and theoretical studies. The methods included evolutionary and memetic (hybrid) algorithms, iterated local search, variable neighborhood search, ant colony optimization, artificial immune systems, hyper-heuristics, and other adaptive approaches. The applications included both traditional domains, such as graph coloring, knapsack, vehicle routing, job-shop scheduling, the p-median, and the orienteering problems; and new(er) domains such as designing deep recurrent neural networks, detecting network community structure, lock scheduling of ships, cloud resource management, the firefighter problem, and AI planning. The theoretical studies involved approximation ratio, runtime, and black-box complexity analyses. The consideration of multiple objectives, dynamic, and noisy environments was also present in a number of articles. This makes the EvoCOP proceedings an important source for current research trends in combinatorial optimization.

EvoCOP	Submitted	Accepted	Acceptance (%)
2015	46	19	41.3
2014	42	20	47.6
2013	50	23	46.0
2012	48	22	45.8
2011	42	22	52.4
2010	69	24	34.8
2009	53	21	39.6
2008	69	24	34.8
2007	81	21	25.9
2006	77	24	31.2
2005	66	24	36.4
2004	86	23	26.7
2003	39	19	48.7
2002	32	18	56.3
2001	31	23	74.2

We would like to express our appreciation to the various persons and institutions making this a successful event. First, we thank the Local Organizers Paolo Burelli from the Aalborg University and Sebastian Risi from the IT University of Copenhagen. We extend our acknowledgment to Pablo García Sánchez from the Universidad de Granada and Mauro Castelli from the Universidade Nova de Lisboa for excellent website and publicity material. We thank Marc Schoenauer from Inria (France) for his continued assistance in providing MyReview conference management system. Thanks are also due to Jennifer Willies and the Institute for Informatics and Digital Innovation at Edinburgh Napier University, UK, for administrative support and event coordination. Finally, we want to thank the National Museum of Denmark at Copenhagen, where the conference was held, and the prominent keynote speakers, Paulien Hogeweg from Utrecht University and Pierre-Yves Oudeyer, Research Director at Inria Paris.

Special thanks also to Carlos Cotta, Peter Cowling, Jens Gottlieb, Jin-Kao Hao, Jano van Hemert, Peter Merz, Martin Middendorf, Günther R. Raidl, and Christian Blum for their hard work and dedication at past editions of EvoCOP, making this one of the reference international events in evolutionary computation and metaheuristics.

April 2015   Gabriela Ochoa
                                                                   Francisco Chicano

Organization

EvoCOP 2015 was organized jointly with EuroGP 2015, EvoMUSART 2015, and EvoApplications 2015.

Organizing Committee

PC Chairs

Gabriela Ochoa	University of Stirling, UK
Francisco Chicano	University of Málaga, Spain

Local Organization

Paolo Burelli	Aalborg University, Denmark
Sebastian Risi	IT University of Copenhagen, Denmark

Publicity Chairs

Pablo García Sánchez	Universidad de Granada, Spain
Mauro Castelli	Universidade Nova de Lisboa, Portugal

EvoCOP Steering Committee

Carlos Cotta	Universidad de Málaga, Spain
Peter Cowling	University of York, UK
Jens Gottlieb	SAP AG, Germany
Jin-Kao Hao	University of Angers, France
Jano van Hemert	University of Edinburgh, UK
Peter Merz	Hannover University of Applied Sciences and Arts, Germany
Martin Middendorf	University of Leipzig, Germany
Günther Raidl	Vienna University of Technology, Austria

Program Committee

Adnan Acan	Eastern Mediterranean University, Turkey
Enrique Alba	Universidad de Málaga, Spain
Mehmet Emin Aydin	University of Bedfordshire, UK
Ruibin Bai	University of Nottingham, UK
Thomas Bartz-Beielstein	Cologne University of Applied Sciences, Germany
Matthieu Basseur	University of Angers, France
Maria J. Blesa	Universitat Politècnica de Catalunya, Spain
Christian Blum	IKERBASQUE and University of the Basque Country, Spain
Sandy Brownlee	University of Stirling, UK

Pedro Castillo	Universidad de Granada, Spain
Francisco Chicano	Universidad de Málaga, Spain
Carlos Coello Coello	CINVESTAV-IPN, Mexico
Peter Cowling	University of York, UK
Karl Doerner	Johannes Kepler University Linz, Austria
Benjamin Doerr	LIX, École Polytechnique, France
Bernd Freisleben	University of Marburg, Germany
Adrien Goeffon	University of Angers, France
Jens Gottlieb	SAP AG, Germany
Walter Gutjahr	University of Vienna, Austria
Jin-Kao Hao	University of Angers, France
Emma Hart	Edinburgh Napier University, UK
Richard F. Hartl	University of Vienna, Austria
Geir Hasle	SINTEF Applied Mathematics, Norway
István Juhos	University of Szeged, Hungary
Graham Kendall	University of Nottingham, UK
Joshua Knowles	University of Manchester, UK
Mario Köppen	Kyushu Institute of Technology, Japan
Frédéric Lardeux	University of Angers, France
Rhyd Lewis	Cardiff University, UK
Arnaud Liefooghe	Université des Sciences et Technologies de Lille, France
José Antonio Lozano	University of the Basque Country, Spain
Gabriel Luque	Universidad de Málaga, Spain
Penousal Machado	University of Coimbra, Portugal
Jorge Maturana	Universidad Austral de Chile, Chile
Barry McCollum	Queen's University Belfast, UK
David Meignan	University of Osnabrück, Germany
Juan Julián Merelo	Universidad de Granada, Spain
Peter Merz	Hannover University of Applied Sciences and Arts, Germany
Martin Middendorf	Universität Leipzig, Germany
Julian Molina	Universidad de Málaga, Spain
Eric Monfroy	University of Nantes, France
Christine L. Mumford	Cardiff University, UK
Nysret Musliu	Vienna University of Technology, Austria
Gabriela Ochoa	University of Stirling, UK
Beatrice Ombuki-Berman	Brock University, Canada
Mario Pavone	University of Catania, Italy
Francisco J.B. Pereira	University of Coimbra, Portugal
Jakob Puchinger	Austrian Institute of Technology, Austria
Günther Raidl	Vienna University of Technology, Austria
Marcus Randall	Bond University, Australia
Marc Reimann	University of Graz, Austria
Eduardo Rodriguez-Tello	Cinvestav - Tamaulipas, Mexico
Peter Ross	Edinburgh Napier University, UK

Contents

A Biased Random-Key Genetic Algorithm for the Cloud Resource Management Problem

Leonard Heilig[1]([✉]), Eduardo Lalla-Ruiz[2], and Stefan Voß[1]

[1] Institute of Information Systems (IWI), University of Hamburg, Hamburg,
Germany
{leonard.heilig,stefan.voss}@uni-hamburg.de
[2] Department of Computer and Systems Engineering, University of La Laguna,
Santa Cruz de Tenerife, Spain
elalla@ull.es

Abstract. Flexible use options and associated cost savings of cloud computing are increasingly attracting the interest from both researchers and practitioners. Since cloud providers offer various cloud services in different forms, there is a large potential of optimizing the selection of those services from the consumer perspective. In this paper, we address the Cloud Resource Management Problem that is a recent optimization problem aimed at reducing the payment cost and the execution time of consumer applications. In the related literature, there is one approach that successfully addresses this problem based on a Greedy Randomized Adaptive Search Procedure. Due to the fact that consumers require fast and high-quality solutions to economically automate cloud resource management and deployment processes, we propose an efficient Biased Random-Key Genetic Algorithm. The computational experiments over a benchmark suite generated based on real cloud market offerings indicate that the performance of our approach outperforms the approaches proposed in the literature.

Keywords: Cloud computing · Cloud resource management · Genetic algorithm · Optimization

1 Introduction

Cloud computing has revolutionized the way information technologies (IT) and related services are offered and consumed by providers and consumers, respectively [1]. In a cloud ecosystem, a cloud service provider (CSP) offers different IT services to multiple consumers, mainly, virtual computing resources (infrastructure as a service – IaaS), development platforms (platform as a service – PaaS), or web-based applications (software as a service – SaaS) [2]. Based on virtualization technology, a multi-tenant model enables that computing resources can be pooled to serve multiple consumers facilitating a cost and energy-efficient utilization of available computing resources assuming that they are allocated efficiently. The consumer may benefit from this by purchasing cloud-based IT services at a lower price. However, the main potentials for consumers are reflected by

© Springer International Publishing Switzerland 2015
G. Ochoa and F. Chicano (Eds.): EvoCOP 2015, LNCS 9026, pp. 1–12, 2015.
DOI: 10.1007/978-3-319-16468-7_1

the flexible and scalable usage options as well as by usage-oriented pricing models associated with on-demand cloud services. As a result, the consumer can adjust the use of cloud services according to the actual and current demand thus avoiding the provision of unused and costly in-house IT infrastructure. There can also be cloud brokers acting on behalf of consumers to negotiate with multiple CSPs in order to bundle demand and achieve better service contracts [3]. As there are several manifestations and options of those cloud services, there are many potentials of optimizing the consumption of cloud services. Both industry and research groups would largely benefit, specifically the research area related to eScience, big data analytics, and high performance computing [4]. While a lot of research is primarily focused on the efficient use of cloud resources from the perspective of CSPs, in particular regarding cost (see, e.g., [5]) and energy-efficiency (see, e.g., [6]), only a few works have considered the consumer perspective in recent years (see, e.g., [7–10]). According to Heilig and Voß [11] and Marston et al. [1], research for understanding and solving consumer-related issues of cloud computing becomes increasingly important to achieve its potentials. As the fast moving cloud market consists of multiple cloud providers that offer multiple resources with various prices and features, it becomes increasingly difficult, in terms of cost and time, for consumers and cloud brokers to choose an appropriate configuration of cloud resources to deploy individual applications. For a general overview on the economics and elasticity challenges of deploying tasks and applications on cloud environments the reader is referred to the work of Suleiman et al. [12].

The previous discussion leads to the development of mathematical models and algorithms for supporting consumers and cloud brokers when choosing and scheduling appropriate configurations of the cloud resources at hand. In this regard, the employed algorithms should assist consumers while optimizing costs and improving the overall performance of service in terms of the consumer specific requirements and preferences. At this point, it is worth highlighting the importance of the computational time within this process. That is, in order to further automate the deployment of the processes, the required computational time of the respective optimization models and algorithms should be considerably low with the aim of allowing users to define cloud resource configurations according to the real-time resource demands of the services. Consequently, fast problem solving techniques, proving satisfactory solutions, are required.

Recently, Coutinho et al. [13] proposed the Cloud Resource Management Problem (CRMP), which is a novel multi-criteria optimization model taking into account both, cost and performance preferences of consumers. The authors present an integer programming (IP) formulation, referred to as CC-IP, which considers resource demands of applications, budget limits, and various resource packages that describe the physical capabilities of associated virtual machine (VM) instances. A Greedy Randomized Adaptive Search Procedure (GRASP) is proposed for solving multiple problem instances. Using CC-IP or GRASP, however, is computationally expensive for multiple problem instances (between minutes and almost an hour for some problem instances). As the computation time plays a decisive role for an efficient decision support allowing the automation of deployment processes, such as discussed in Heilig *et al.* [14], there is an urgent need of providing a faster algorithm for solving the CRMP.

The goals of this paper are, on the one hand, to propose a Biased Random-Key Genetic Algorithm (BRKGA) approach for solving the CRMP, referred to as BRKGA-CC. On the other hand, we aim to develop a fast algorithm that provides high-quality solutions in terms of objective function value, that supports either, consumers and cloud brokers, when addressing the packages selection. It should be noted that fast algorithms may be desirable from the consumer viewpoint, as such an algorithm could be executed on a standard user computer without requiring expensive high performance computing. For evaluating its performance, we undertake an extensive experimentation over the benchmark suite proposed for this problem in Coutinho *et al.* [13]. In a nutshell, our evolutionary approach noticeably outperforms recent approaches within small computation times in the millisecond range, on average. This represents an important step towards real-time decision support for deploying applications in the cloud, which is not only essential for individual consumers, but also for cloud brokers as discussed in Lucas-Simarro *et al.* [15] and Tordsson *et al.* [16].

The remainder of this paper is organized as follows. In Sect. 2, the CRMP is described. Section 3 presents the proposed BRKGA-CC for solving this problem. Computational results are given in Sect. 4. Finally, conclusions and some lines for future work are stated in Sect. 5.

2 Cloud Resource Management Problem

The CRMP is proposed by Coutinho *et al.* [13] as a cloud computing optimization problem oriented towards consumers and cloud brokers. That is, in the CRPM, we specifically focus on IaaS enabling consumers to select and utilize resizable computational resources to run their applications, after the consumer has decided which CSP will be used (for example, *Amazon Web Services*, *Google Cloud Platform*, etc.). In this problem, the main difficulty lies in finding an appropriate combination of cloud resources that satisfy application requirements and specific consumer requirements, which may be predefined by budget and performance constraints. Application requirements may include the minimum number of Giga floating point operations (Gflop) to execute the application, total memory requirements (in Gigabytes), and maximum hard disk capacity (in Gigabytes). As depicted in Fig. 1, a CSP offers different packages of VM instances. A package describes the specific attributes and a respective price. The attributes include processor capacity (measured in Gflop), memory capacity, platform architecture (32-bit/64-bit), and hard disk capacity. Not surprisingly, the price for a package with high computational resource capacities is higher than for lower capacities. Vice versa, the performance of packages with high computational resource capacities is higher thus enabling an application to be executed faster. Consequently, a conflict of objectives is likely to occur. To handle cost-performance trade-offs, the consumer must specify the weight of such factors. Based on the multicriteria objective function, an optimal selection of cloud resources must be determined, considering application and consumer requirements. Finally, the consumer utilizes the selected resources by deploying its applications.

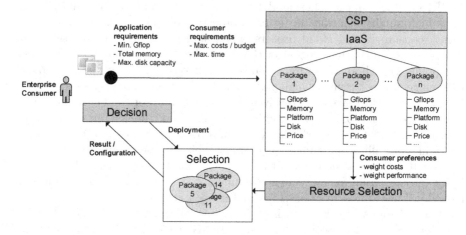

Fig. 1. Cloud resource management problem

2.1 Mathematical Formulation

In the following, with the aim of making this paper self-contained, we include the mathematical formulation of the CRMP as presented in Coutinho *et al.* [13].

For a consumer, we define both, a set of application requirements and a set of consumer requirements. The set of application requirements consists of a minimum demand of processing capacity of G_f Gflop, memory capacity of m_p, and hard disk capacity of D_S. On the other hand, the consumer requirements consist of the maximum payment cost C_M, expressing the consumer's willingness to pay (*i.e.*, budget limit), as well as an execution time limit T_M associated with the deployment of an application in the cloud of a CSP. Considering the CSP's offer, let P denote the package types of a VM offered during a set of time periods $t \in T = \{1, 2, ..., T_M\}$. Each package type $p \in P$ has associated computational resource capacities expressed by the processing capacity of g_p Gflop per period t (Gflop(t)), memory capacity m_p, and hard disk capacity d_p. The cost c per period for running a VM of package type p is denoted as c_p, which corresponds to the price of package type p charged by the CSP in each period. Moreover, a maximum limit of N_M packages that can be purchased at each time period is defined according to common policies used by CSPs. A binary variable x_{pit} is modeled for each $p \in P$, $i \in \{1, 2, ..., N_M\}$ and $t \in T$, such that $x_{pit} = 1$ if package i of type p is consumed at time t and $x_{pit} = 0$, otherwise. The variable t_m denotes the last time period in which a package was purchased by the consumer.

For a detailed description the reader is referred to Coutinho *et al.* [13].

$$\text{(CC-IP) } minimize \left(\alpha_1 \sum_{p \in P} \sum_{i=1}^{N_M} \sum_{t \in T} c_p x_{pit} + \alpha_2 t_m \right) \tag{1}$$

subject to:

$$\sum_{p \in P} \sum_{i=1}^{N_M} \sum_{t \in T} c_p x_{pit} \leq C_M \tag{2}$$

$$\sum_{p \in P} \sum_{i=1}^{N_M} d_p x_{xip} \geq D_S \ x_{p'i't}, \forall t \in T, \forall p' \in P, \forall i' \in \{1, ..., N_M\} \tag{3}$$

$$\sum_{p \in P} \sum_{i=1}^{N_M} m_p x_{xip} \geq M_C \ x_{p'i't}, \forall t \in T, \forall p' \in P, \forall i' \in \{1, ..., N_M\} \tag{4}$$

$$\sum_{p \in P} \sum_{i=1}^{N_M} \sum_{t \in T} g_p x_{pit} \leq G_f \tag{5}$$

$$\sum_{p \in P} \sum_{i=1}^{N_M} x_{pit} \leq N_M, \forall t \in T \tag{6}$$

$$t_m \geq t \ x_{pit}, \forall t \in T, \forall p \in P, \forall i \in \{1, ..., N_M\} \tag{7}$$

$$x_{pit+1} \leq x_{pit}, \forall t \in T, \forall p \in P, \forall i \in \{1, ..., N_M\} \tag{8}$$

$$x_{pi+1t} \leq x_{pit}, \forall t \in T, \forall p \in P, \forall i \in \{1, ..., N_M - 1\} \tag{9}$$

$$x_{pit} \in \{0,1\}, \forall t \in T, \forall p \in P, \forall i \in \{1, ..., N_M\}, t_m \in \mathbb{Z} \tag{10}$$

where $(\alpha_1 + \alpha_2) = 1$.

The objective function of the CC-IP model (1) seeks both the minimization of costs and execution time (*i.e.*, total purchased time). As these objectives are conflicting, there is no common alternative for providing an optimal solution for both objectives simultaneously. To handle the cost-performance trade-off, the target weights α_1 and α_2 enable the consumer to specify the importance of each objective. For values of α_1 close to 1, the consumer prefers low-budget solutions at an expense of larger total execution time. For values of α_2 close to 1, correspondingly, the consumer prefers a short execution time at an expense of higher total costs. Constraints (2) ensure that the cost for purchasing packages do not exceed the budget limit of the consumer. Constraints (3) and (4) state the minimum application resource demands in terms of memory and hard disk capacity at each time period. Equally, constraints (5) guarantee that the purchased processing capacity is large enough to satisfy the overall demand. Constraints (6) ensure that the maximum limit of packages per time period is not exceeded. Constraints (7) define the properties of variable t_m, while constraints (8) guarantee that there are no gaps of time in a feasible solution. In this regard, a package that is selected at time $t + 1$, is also selected at time t. Constraints (9) establish an order between packages and eliminates symmetries. Constraints (10) define the binary and integer variables x_{pit} and t_m.

3 Biased Random-Key Genetic Algorithm

The Biased Random Key Genetic Algorithm, BRKGA, (see [17,18]) is presented
as a variation of the Random Key Genetic Algorithm (RKGA) proposed by
Bean [19]. It differs in the way the crossover is performed. That is, for generating
the offspring, BRKGA selects one parent from an elite population and the other
one from the rest of the population. Moreover, in the crossover process, for giving
more probability to the elite parent genes, a biased coin favouring the elite parent
is tossed, so the child would have more probability of inhering the keys of its elite
parent. This strategy implies an implicit learning from the best solutions along
the generations. In the BRKGA, we have a fixed-size population, Pop, consisting
of $|Pop|$ individuals, where each individual is a sequence of randomly generated
numbers (random keys) in the real interval $[0, 1]$. Through a decoder, a vector of
random keys r is translated to a solution of the optimization problem at hand
with a fitness value $f(r)$ for that vector. At each iteration of the algorithm, the
population is partitioned into a smaller set Pop_e of elite individuals and a larger
set Pop_c with the remaining individuals of Pop. The evolutionary process within
BRKGA is as follow. First, all elite individuals are copied, without change, to
the population of the next generation. Then, a set Pop_m of mutant individuals,
generated in the same way as an individual of the initial population is inserted
into the population of the next generation. Finally, the rest of the population is
filled with the offspring obtained through parameterized uniform crossover with
an inheritance probability ς. This crossover follows an elitist strategy, *i.e.*, one
parent, rka, from Pop_e and one, rkb, from Pop_c are selected at random. For a
more detailed explanation the reader is referred to [17,18].

3.1 BRKGA Application for the CRMP

Initial Population. The initial population is composed of $|Pop|$ chromosomes,
each randomly generated.

Coding. The solutions for the CRMP are encoded with an n-dimensional array
$R = (r_1, r_2, ..., r_{|2 \cdot P|})$, where $|P|$ is the number of types of packages. Each component r_i $(i = 1, ..., n)$ is a real number in the interval $[0, 1]$.

Decoding. In the decoding process, a random key vector is mapped in the
solution space of the CRMP. In this regard, through each random key vector,
R, we determine the preference and quantity of packages used. For this purpose,
the random key vector is divided into two parts:

1. Firstly, we order increasingly the first $|P|$ random keys of R, from the leftmost
   gene up to the $|P|$-th rightmost gene, giving rise to a sorted random key vector
   denoted as $R_1 = (r'_1, r'_2, ..., r'_{|P|})$, where $r'_i \leq r'_{i+1}$, $i = 1, 2, ..., |P| - 1$. The
   position, k, that each gene $r'_i \in R_1$ occupies in R determines the k-type of
   package preference used when determining the quantity of packages in the
   next step. Thus, for example, the position of the gene with lowest value in

the disordered sequence, R, establishes the package type with the highest preference when assigning the number of packages in the next point. The preference array obtained in this step is denoted as $O = (o_1, o_2, ..., o_{|P|})$, where each o_i, $i = 1, 2, ..., |P|$, is a type of package.

2. The interval $[0, 1]$ is equally divided by variable δ, which is defined as the number of maximum packages, N_M, minus the number of packages already assigned in the decodification process. Hence, initially $\delta = N_M$, which results in N_M subintervals corresponding to the possible quantity of packages we can assign. Once we have the δ intervals, we traverse R for $i = |P| + 1, ..., |2 \cdot P|$, if the r_i value is within an interval, then the quantity represented by that interval is selected to be assigned to the package type established by $o_{i-|P|}$. Note that this preference array, O, is determined in the previous step. Once that happens, the δ value is updated. The process of determining the quantities may also finish when $\delta = 0$.

3. Once the package types and quantities are determined, we have a *profile* per each time step. This profile is repeated until all the G_f required are provided. Note that this repetition may cause a surplus of Gflops. To address this, we use an algorithm for reducing that surplus and, therefore, the total cost. The algorithm reduces the Gflops as much as possible by considering the unit reduction of the packages. This reduction considers the constraints established for the problem.

Crossover. Crossover is carried out by combining the genes of two chromosomes, one randomly chosen from P_e and one from the rest of the population. The combination strategy used is the parameterized uniform crossover, where an inheritance probability ς has been selected.

Mutation. The mutation is carried out in order to prevent premature convergence and expand the search space. Unlike genetic recombination, where mutation is performed by modifying genes of certain chromosomes in the population, the mutation strategy used in this work is based on the criterion of immigration (see [19]). That is, at each generation a small number $|Pop_m|$ of individuals are randomly generated and included in the new population.

Reproduction. The reproduction strategy used in this algorithm is based on *elitism*. That is, the best individuals of the population Pop_e are copied from one generation to the next one. The advantage of this strategy is that it maintains the best solutions while the population is being improved and ensures a constant number of good individuals for mating.

4 Computational Results

This section is devoted to assess the performance of our proposed algorithm. In doing so, we have tested it over a set of problem instances based on real cloud

Table 1. Instance description of Coutinho *et al.* [13]

Instance	Memory (GB)	Storage (GB)	Gflop (t)	Time (hr.)	Max. Packages	Cost ($)
nug22-sbb	77	51	5,067,533	12	20	343
nug24-sbb	154	103	14,741,914	48	20	998
nug25-sbb	214	142	28,792,800	60	20	1,950
nug28-sbb	528	352	67,720,666	72	20	4,586
nug30-sbb	918	612	120,929,760	84	20	8,190
nug22-cbb	77	51	20,270,131	12	20	343
nug24-cbb	154	154	88,451,482	72	20	1,498
nug25-cbb	214	214	230,342,400	120	20	3,900
nug28-cbb	528	528	541,765,325	144	20	9,173
nug30-cbb	918	918	967,438,080	168	20	16,380
mod-gen	4	2	3,317,760	24	20	100
raxml	3	2	3,317,760	24	20	100
segemehl	64	600	28,748,390	4	20	192
cms-1000	1500	20	216,000,000	24	30	1,728
cms-1250	1875	25	270,000,000	24	40	2,304
cms-1500	2250	30	324,000,000	24	45	2,592

market offerings proposed in Coutinho *et al.* [13]. These instances include different requirement sets on Gflop, memory, and hard disk capacity of five real applications [13]: a branch-and-bound algorithm for solving the Quadratic Assignment Problem (QAP) [20], three algorithms that tackle the manipulation of a biological sequence problem (RAXML [21], ModelGenerator [22] and Segemehl [23]), and a typical user analysis job for the CMS experiment [24]. Each instance further includes a maximum allowed execution time and payment cost, defined by the consumer. Moreover, two sets of VM packages available in two commercial clouds, namely *Amazon EC2* and *Google Compute Engine*, are used to define available package types for each instance as well as a maximum limit of packages that can be purchased in each period of time, defined by the respective CSP. Having the data of different CSPs, a comparison between multiple CSPs is possible. Table 1 describes the used instances.

The proposed optimization technique has been implemented in JAVA and executed on a computer equipped with an Intel 3.16 GHz and 4 GB of RAM. By preliminary experiments, we identified the following parameters. A population, *Pop*, consisting of 100 individuals is used, within which an elite population, Pop_e, of 10 and a mutation population, Pop_m, of 10 individuals are considered. The inheritance probability, ς, is set to 0.8. For each instance, we have performed 20 executions. Table 2 illustrates the results provided by CC-IP [13], GRASP-CC [13] and our approach, BRKGA-CC for the instances proposed by [13] using the

Table 2. Computational results of GraspCC [13], CC-IP [13], and BRKGA-CC with $\alpha_1 = 0.5$ and $\alpha_2 = 0.5$. * indicate that CPLEX reached time limit. ** indicate that the reported value is not feasible.

Instances	CC-IP		GraspCC			BRKGA-CC		
	Obj.	t(s.)	Obj.	Gap (%)	t(s.)	Obj.	Gap (%)	t(s.)
nug22-sbb_am	2.75	0.59	2.75	0.00	0.01	2.75	0.00	0.03
nug24-sbb_am	6.55	6.79	6.55	0.00	0.02	6.55	0.00	0.05
nug25-sbb_am	9.05	10.98	9.05	0.00	0.05	9.05	0.00	0.07
nug28-sbb_am	20.90	27.23	20.90	0.00	0.07	20.90	0.00	0.04
nug30-sbb_am	68.25	32.42	68.25	0.00	0.08	68.25	0.00	0.07
nug22-cbb_am	6.40	0.77	6.40	0.00	0.03	6.40	0.00	0.02
nug24-cbb_am	24.10	5.81	24.10	0.00	0.17	24.10	0.00	0.03
nug25-cbb_am	61.35	12.01	61.35	0.00	0.91	61.35	0.00	0.02
nug28-cbb_am	18.95	21.20	18.95	0.00	0.07	18.95	0.00	0.03
nug30-cbb_am	528.85	118.42	528.85	0.00	3.63	528.85	0.00	0.04
mod-gen_am	1.55	0.50	1.55	0.00	0.00	1.55	0.00	0.06
raxml_am	1.55	0.48	1.55	0.00	0.00	1.55	0.00	0.08
segemehl_am	8.90	0.26	8.90	0.00	0.07	8.90	0.00	0.06
cms-1000_am	139.60	31.82	139.60	0.00	0.18	139.60	0.00	0.17
cms-1250_am	172.10	39.50	172.10	0.00	0.53	172.10	0.00	0.13
cms-1500_am	207.20	32.83	207.20	0.00	0.53	207.20	0.00	0.16
nug22-sbb_go	2.94	77.72	2.95	0.34	5.36	2.94	0.00	0.14
nug24-sbb_go**	7.98**	9194.37	7.98**	0.00	20.74	8.48	6.27	0.16
nug25-sbb_go	16.27	28081.33	16.27	0.00	38.90	16.27	0.00	0.29
nug28-sbb_go	46.72	35278.98	46.72	0.00	133.66	46.72	0.00	0.20
nug30-sbb_go	106.64	73656.27	106.64	0.00	42.98	106.64	0.00	1.18
nug22-cbb_go	10.28	246.79	10.28	0.00	17.97	10.28	0.00	0.19
nug24-cbb_go	45.46	19695.32	45.46	0.00	177.53	45.46	0.00	0.28
nug25-cbb_go	124.50	64758.31	124.50	0.00	1186.05	124.50	0.00	0.18
nug28-cbb_go*	37.82	86400.00*	37.87	0.12	60.00	37.82	0.00	0.21
nug30-cbb_go	827.11	5085.60	827.11	0.00	2987.79	827.11	0.00	0.74
mod-gen_go	2.08	32.71	2.08	0.00	2.97	2.08	0.00	0.22
raxml_go	2.08	29.34	2.08	0.00	2.91	2.08	0.00	0.29
segemehl_go	14.61	23.52	14.61	0.00	52.40	14.61	0.00	0.29
cms-1000_go	198.85	4458.83	198.85	0.00	27.56	198.85	0.00	2.31
cms-1250_go	230.76	34210.73	230.76	0.00	153.13	230.76	0.00	0.87
cms-1500_go*	293.43	86400.00*	293.43	0.00	53.27	293.23	-0.07	1.84
Average	101.42	13999.11	101.43		155.30	101.43		0.33

same α values, *i.e.*, $\alpha_1 = 0.5$ and $\alpha_2 = 0.5$. Both, CC-IP and GRASP-CC, as indicated in [13] were executed on a computer equipped with an Intel Core i7 3.4 GHz and 12 GB of RAM. For each instance, we report the best objective value (Obj.) and the running time of each approach (t(s.)) measured in seconds. Finally, for each approximate approach we include the relative error (Gap(%)) based on the values reported by CC-IP.

As can be checked in Table 2, BRKGA-CC outperforms CC-IP and GraspCC in terms of running time. This characteristic is relevant if we take into account that the required computational times should be considerably low with the aim of allowing an automatic selection and deployment of cloud resources according to the real-time resource demands of applications. In this sense, low-computational times as the ones reported by BRKGA-CC allow consumers and cloud brokers to test different profiles or to evaluate more possibilities when several cloud providers have to be selected. Furthermore, concerning the quality of the solutions in terms of the objective function value, we can highlight that the BRKGA-CC presents a competitive performance, allowing the improvement of solutions in comparison with GraspCC and provide a new best value for an instance (cms-1500_go) where CPLEX reaches a time limit.

Problem Instance Nug-24-sbb_go. As provided in Table 2, the objective values reported by CC-IP and GraspCC are not feasible in terms of the mathematical formulation. As shown in Table 1, we are required to fulfill 14,741,914 Gflop and we have a maximum number of 20 packages per time period. Moreover, as included in the instance, the maximum Gflop that a package can contribute is $(166.4 \cdot 3600 = 599,040)$. Hence, if 20 of those packages are selected, then, at most 11,980,800 Gflop can be obtained in each time step. Therefore, we will need more than $t_m = 1$ to fulfill the amount of Gflop required.

5   Conclusions

In this paper, we propose a Biased Random-Key Genetic Algorithm, BRKGA-CC, for solving the CRMP. This recent cloud computing optimization problem is aimed at finding an appropriate configuration while satisfying application demands, in particular if respective decision makers have multiple, possibly conflicting objectives. The development of algorithms for providing decision support when choosing sets of VM packages (offered by cloud providers) becomes increasingly important. In this sense, the algorithms should assist consumers while optimizing costs and improving the overall performance of service in terms of consumer specific requirements (e.g., budget constraints, maximum allowed execution time) and preferences (e.g., low-budget versus high-performance computations). The required computational time of the respective optimization models and algorithms should be considerably low with the aim of allowing an automatic selection and deployment of cloud resources according to the real-time resource demands of applications. In return, the consumer is able to better utilize the flexible use options and attractive prices of resources in the cloud over time in order to achieve reduced operational costs and a higher quality of service for deployed applications.

The computational experiments reported by BRKGA-CC over a benchmark suite generated based on real data indicate that the performance of our approach outperforms the approaches that were recently proposed in the literature in terms of computational time and quality. Our algorithm provides high quality solutions, compared to recent approaches, and achieves computation times in the millisecond range for all instances. As a result, thanks to the CPU-time advantage of BRKGA-CC, it enables real-time decision support for deploying applications in the cloud both important for cloud consumers and brokers. Moreover, it can be indicated that BRKGA-CC is a suitable algorithm for being used either isolatedly or within integration schemes when addressing cloud resource management related problems.

For future work, we intend to extend the model with brokers acting on behalf of multiple consumers with individual preferences and constraints. In this scenario, the broker selects sets of configurations offered by multiple CSPs in order to satisfy individual consumer requirements.

Acknowledgements. This work has been partially funded by the European Regional Development Fund, the Spanish Ministry of Economy and Competitiveness (project TIN2012-32608). Eduardo Lalla-Ruiz thanks the Canary Government for the financial support he receives through his doctoral grant. We thank the authors of [13] for providing their data.

References

1. Marston, S., Li, Z., Bandyopadhyay, S., Zhang, J., Ghalsasi, A.: Cloud computing - the business perspective. Decis. Support. Syst. **51**(1), 176–189 (2011)
2. Vaquero, L.M., Rodero-Merino, L., Caceres, J., Lindner, M.: A break in the clouds: towards a cloud definition. ACM SIGCOMM Comput. Commun. Rev. **39**(1), 50–55 (2009)
3. Buyya, R., Yeo, C.S., Venugopal, S., Broberg, J., Brandic, I.: Cloud computing and emerging IT platforms: vision, hype, and reality for delivering computing as the 5th utility. Future Gener. Comput. Syst. **25**(6), 599–616 (2009)
4. Hoffa, C., Mehta, G., Freeman, T., Deelman, E., Keahey, K., Berriman, B., Good, J.: On the use of cloud computing for scientific workflows. In: 4th International Conference on eScience (eScience 2008), pp. 640–645. IEEE (2008)
5. Van den Bossche, R., Vanmechelen, K., Broeckhove, J.: Cost-optimal scheduling in hybrid IaaS clouds for deadline constrained workloads. In: 3rd IEEE International Conference on Cloud Computing (CLOUD 2010), pp. 228–235. IEEE (2010)
6. Beloglazov, A., Abawajy, J., Buyya, R.: Energy-aware resource allocation heuristics for efficient management of data centers for cloud computing. Future Gener. Comput. Syst. **28**(5), 755–768 (2012)
7. Gutierrez-Garcia, J.O., Sim, K.M.: A family of heuristics for agent-based elastic cloud bag-of-tasks concurrent scheduling. Future Gener. Comput. Syst. **29**(7), 1682–1699 (2013)
8. Mao, M., Li, J., Humphrey, M.: Cloud auto-scaling with deadline and budget constraints. In: 11th IEEE/ACM International Conference on Grid Computing (GRID 2010), pp. 41–48. IEEE (2010)

9. Netjinda, N., Achalakul, T., Sirinaovakul, B.: Cloud provisioning for workflow application with deadline using discrete PSO. ECTI Trans. Comput. Inf. Technol. **7**(1), 43–51 (2013)
10. Oprescu, A., Kielmann, T.: Bag-of-tasks scheduling under budget constraints. In: IEEE Second International Conference on Cloud Computing Technology and Science (CloudCom 2010), pp. 351–359. IEEE (2010)
11. Heilig, L., Voß, S.: A scientometric analysis of cloud computing literature. IEEE Trans. Cloud Comput. **2**(3), 266–278 (2014)
12. Suleiman, B., Sakr, S., Jeffery, R., Liu, A.: On understanding the economics and elasticity challenges of deploying business applications on public cloud infrastructure. J. Internet Serv. Appl. **3**(2), 173–193 (2012)
13. Coutinho, R.d.C., Drummond, L.M.A., Frota, Y.: Optimization of a cloud resource management problem from a consumer perspective. In: an Mey, D., et al. (eds.) Euro-Par 2013. LNCS, vol. 8374, pp. 218–227. Springer, Heidelberg (2014)
14. Heilig, L., Voß, S., Wulfken, L.: Building clouds: An integrative approach for an automated deployment of elastic cloud services. In: Chang, V., Walters, R., Wills, G. (eds.) Delivery and Adoption of Cloud Computing Services in Contemporary Organizations. IGI Global (2015, to appear)
15. Lucas-Simarro, J.L., Moreno-Vozmediano, R., Montero, R.S., Llorente, I.M.: Scheduling strategies for optimal service deployment across multiple clouds. Future Gener. Comput. Syst. **29**(6), 1431–1441 (2013)
16. Tordsson, J., Montero, R.S., Moreno-Vozmediano, R., Llorente, I.M.: Cloud brokering mechanisms for optimized placement of virtual machines across multiple providers. Future Gener. Comput. Syst. **28**(2), 358–367 (2012)
17. Ericsson, M., Resende, M., Pardalos, P.M.: A genetic algorithm for the weight setting problem in OSPF routing. J. Comb. Optim. **6**, 299–333 (2002)
18. Gonçalves, J.F., Resende, M.: Biased random-key genetic algorithms for combinatorial optimization. J. Heuristics **17**, 487–525 (2011)
19. Bean, J.C.: Genetic algorithms and random keys for sequencing and optimization. RSA J. Comput. **6**, 154–160 (1994)
20. Gonçalves, J.F., Drummond, L.M., Pessoa, A.A., Hahn, P.: Improving lower bounds for the quadratic assignment problem by applying a distributed dual ascent algorithm. arXiv preprint arXiv:1304.0267 (2013)
21. Stamatakis, A.: RAxML-VI-HPC: maximum likelihood-based phylogenetic analyses with thousands of taxa and mixed models. Bioinformatics **22**(21), 2688–2690 (2006)
22. Keane, T.M., Creevey, C.J., Pentony, M.M., Naughton, T.J., McInerney, J.O.: Assessment of methods for amino acid matrix selection and their use on empirical data shows that ad hoc assumptions for choice of matrix are not justified. BMC Evol. Biol. **6**(1), 1–17 (2006)
23. Hoffmann, S., Otto, C., Kurtz, S., Sharma, C.M., Khaitovich, P., Vogel, J., Stadler, P.F., Hackermüller, J.: Fast mapping of short sequences with mismatches, insertions and deletions using index structures. PLoS Comput. Biol. **5**(9), 1–10 (2009)
24. Chatrchyan, S., Hmayakyan, G., Khachatryan, V., Sirunyan, A., Adam, W., Bauer, T., Bergauer, T., Bergauer, H., Dragicevic, M., Erö, J., et al.: The CMS experiment at the CERN LHC. J. Instrum. **3**(8) (2008)

A Computational Comparison of Different Algorithms for Very Large p-median Problems

Pascal Rebreyend[1][(✉)], Laurent Lemarchand[2], and Reinhardt Euler[2]

[1] School of Technology and Business Studies, Dalarna University, Falun, Sweden
prb@du.se
[2] Lab-STICC/UBO, Université Européenne de Bretagne, Brest, France
{laurent.lemarchand,reinhardt.euler}@univ-brest.fr

Abstract. In this paper, we propose a new method for solving large scale p-median problem instances based on real data. We compare different approaches in terms of runtime, memory footprint and quality of solutions obtained. In order to test the different methods on real data, we introduce a new benchmark for the p-median problem based on real Swedish data. Because of the size of the problem addressed, up to 1938 candidate nodes, a number of algorithms, both exact and heuristic, are considered. We also propose an improved hybrid version of a genetic algorithm called impGA. Experiments show that impGA behaves as well as other methods for the standard set of medium-size problems taken from Beasley's benchmark, but produces comparatively good results in terms of quality, runtime and memory footprint on our specific benchmark based on real Swedish data.

1 Introduction

Facility location problems consider a set of demand points to be served from a set of possible locations. Solution quality takes into account costs for associating demand points to locations and also costs of choosing a particular location. In the p-median version, the latter is not considered, i.e., we can describe the p-median problem as finding a set of p facilities such that the sum of distances between demand points and the closest facility is minimized. The p-median problem has been introduced by Hakimi [15] who describes its basic properties.

Previous approaches to the p-median problem [1,11,14] have included numerical tests using Beasley's benchmark [7]. The largest graph of this benchmark has 900 candidate nodes on which we can locate facilities. Graphs are generated with a uniform distribution of the demand, which is not representing accurately a true problem since the population in most countries and therefore the demand is not uniformly distributed due to the presence of natural factors like islands, lakes, mountains, rivers,..., and the concentration areas related to urbanization. Further, to represent most regional or national location problems with just 900 candidate nodes is rather limited and implies a high degree of problem simplification together with a loss of information accuracy. Another set of abstract graphs has been used by Avella [4], with the same drawback of uniform distribution of the demand.

© Springer International Publishing Switzerland 2015
G. Ochoa and F. Chicano (Eds.): EvoCOP 2015, LNCS 9026, pp. 13–24, 2015.
DOI: 10.1007/978-3-319-16468-7_2

Due to these two limitations of the Beasley benchmark, one of our interest in this paper is to introduce and use a new set of problem instances based on real data. This new set is based on Swedish real data. Sweden is a good candidate to test methods for the p-median problem since the distribution of population is not uniform as we can see in Fig. 1. The size of the country $(449, 964 \, \text{km}^2)$ is big enough to test algorithms designed for country-related problems. Sweden exhibits also some particular characteristics which make a clear difference between an abstract graph and those obtained from real data such as the presence of natural barriers i.e., lakes or mountains. The non-uniform distribution of the population is also taken into account by distinguishing between demand nodes and candidate nodes. Demand nodes indicate where people are living while candidate nodes are possible locations for a facility. The number of facilities p will vary from 10 to 100 in our cases to cover different practical p-median problems such as locating universities $(p = 10)$, courts, public hospitals $(p = 100)$,.... Locating hospitals is the practical problem we focus on, without limitation on the capacity of a facility. This corresponds to most of the computational experiments done so far [22].

Some tests on larger graphs have been carried out by Avella [4] using the Birch set of abstract graphs (with up to 89 000 nodes). Birch graphs have similar to those of Beasley a uniform distribution of the demand but points are grouped into clusters. Real graphs with up to 67,000 nodes have been used by Rebreyend et al. [21] to investigate effects of the quality of different road networks on the p-median problem. But in their case only one heuristic method was tested and the number of candidate nodes to locate on was high in comparison to the geographical area as was the number of nodes representing locations of population, since data represent only a single Swedish province. The authors of [21] conclude that fewer nodes lead to better results also because they are using an approximate method.

Sweden has an asymmetrical distribution of the population and natural barriers are spread all over the country. Therefore, we need to use the road distances since Euclidian distances may lead to poor results [9].

The p-median problem is described in the next section. Section 3 presents previous work and algorithms related to the p-median problem. A detailed explanation on the algorithms we have used is given in Sect. 4. Section 5 describes our new algorithm. Section 6 presents the data used. Results in terms of quality, runtime and memory footprint for the tested algorithms in Sects. 7, and Sect. 8 concludes the paper.

2 The p-median Problem

In the rest of this paper, the following terms are used:

- N for the number of candidate nodes (number of possible locations for a facility),
- D for the number of demand nodes,
- p for the number of facilities to allocate.

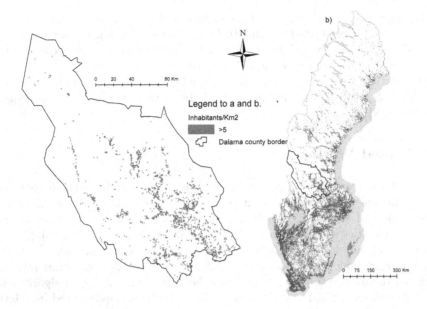

Fig. 1. Distribution of two populations, Dalecarlia province on the left, Sweden on the right

A common formulation of the p-median problem due to Revelle and Swain [23] is the following:

$$\text{minimize} \sum_{i=1}^{D} \sum_{j=1}^{N} h_i d_{ij} Y_{ij} \tag{1}$$

$$\text{subject to} \sum_{j=1}^{N} Y_{ij} = 1 \qquad \forall i, 1 \le i \le D \tag{2}$$

$$\sum_{j=1}^{N} X_j = p \tag{3}$$

$$Y_{ij} - X_j \le 0 \qquad \forall i, j, \, 1 \le i \le D, \, 1 \le j \le N \tag{4}$$

$$X_j \in \{0, 1\} \qquad \forall j, \, 1 \le j \le N \tag{5}$$

$$Y_{ij} \in \{0, 1\} \qquad \forall i, j, \, 1 \le i \le D, \, 1 \le j \le N \tag{6}$$

where

- h_i is the weight of the demand of customer i,
- d_{ij} is the distance between customer i and facility j,
- X_i is a decision variable, indicating whether facility i is selected or not,
- Y_{ij} is a decision variable, indicating whether customer i is served by facility j or not,
- p is the number of facilities to be selected.

Equation (2) ensures that a customer is served by exactly one facility. The number of facilities is fixed by Eq. (3). Constraint set (4) reflects the objective that demand points are assigned only to a selected facility. Binary decision variables X and Y are defined by Eqs. (5) and (6).

At least one optimal location exists if the facilities are located on the nodes of the graph only [15]. Kariv and Hakimi [16] have shown that the p-median problem is NP-Hard.

3 Related Work

Reese [22] has recently published a bibliography on the main methods used for solving the p-median problem. A survey of metaheuristics has been published by Mladenovic [20]. The earliest solution techniques mentioned are enumeration-based or heuristics such as vertex-substitution [3, 18]. Simulated Annealing (SA) based approaches have also been applied to the p-median problem [1, 9, 20]. Genetic algorithms (GA) have been used in [2, 11]. Some approaches rely on hardware to improve runtimes, and thus enlarge the applicability of algorithms. Parallel versions of GA-based approaches have also been implemented [8]. More specifically, GPU-based implementations of the Vertex substitution heuristic [17] or of the Volume algorithm onto multi-core systems [14] can lead to impressive speedups. Exact methods like 0-1 programming have also been proposed [5].

In this paper we focus on sequential algorithms that are suitable for very large problem instances. A detailed description will be given in the next section.

4 Tested Algorithms

4.1 CPlex

The p-median problem can be formulated as a 0-1 programming problem (BP) and then be solved by a Mixed Integer Problem (MIP) solver, using a branch and cut approach. In our tests, we have used the CPlex software from IBM (version 12.6, Linux 64 bits) to test the BP approach. Some parameters of the solver have been tuned in order to adapt CPlex to work on large problem instances, i.e., removing default computation time limits, allowing intermediate data storage, and tuning branch & cut search tree strategies according to [13]. In the following we will refer to this implementation as CPLEX.

4.2 Volume

Barahona and Anbil have given in [6] a description of the Volume algorithm. This algorithm solves Mixed Integer Programs (MIP) by working on the dual of a linear problem using the sub-gradient method [14]. At each iteration approximations of the primal variable values are computed in addition. By working both on the primal and dual problem, the Volume algorithm computes lower and upper bounds for a given problem instance very efficiently, thanks to the subgradient method. The Volume algorithm has been successfully tested on the p-median problem by specializing it to the associated LP-relaxation. The relaxed solution is then exploited by fast heuristics to compute an integral solution of the original p-median problem [10, 14].

The version of the Volume algorithm we used[1] to solve the p-median problem is based on the formulation given in Sect. 2, except that constraint (4) is replaced by the following [10]:

$$\sum_{i \neq j} Y_{ij} + X_j = 1, \forall j, 1 \leq j \leq N \qquad (7)$$

$$Y_{ij} \leq X_j, \forall i, j, 1 \leq i \leq D, 1 \leq j \leq N \qquad (8)$$

Equation (7) indicates that either a node is selected, or it is connected to one candidate. Equation 8 is identical to constraint (4). As a first step for solving a p-median problem instance, the LP-relaxation of the dual problem is formulated, and the heuristic of Bourges-Cleraux [10] is used to find a feasible integer solution from the vector of real numbers found: the p highest values of this vector are selected as the set of locations used. The second step is to select for each customer the closest facility. This is done by sorting edges of the graph according to their distance, from the shortest to the highest and going through them. The next step is to go through all edges in this order. If an edge connects a selected location to a customer with no location assigned to, the corresponding location is assigned to this customer. Altogether, the complexity of this heuristic is $O(m \log m)$ if m is the number of edges since a sort on all edges is done. Default values have been used for the different parameters.

4.3   Simulated Annealing

Al-Khedhairi presents a general version of a Simulated Annealing (SA) algorithm for the p-median problem [1]. Carling et al. have later used a similar approach to solve the p-median problem in the real context of a Swedish province [9, 21].

In this paper, we use Carling et al.'s algorithm. The starting point is a random solution of the p-median problem instance. The neighbourhood of a solution s is defined as the set of all solutions s' in which one of the selected nodes of s has been replaced by another candidate node. All the candidate nodes have the same probability to be chosen.

The initial temperature of the SA is fixed to $400°$. At every iteration the temperature is multiplied by 0.95. A main concern with simulated annealing is the risk to get stuck at a frozen state. To detect such a situation and reheat the system, we proceed as follows: if 10 consecutive iterations do not result in any improvement, the following formula is applied to modify the temperature t: $t = t * 3^\beta$. The initial value of β has been set to 0.5 after some experiments. If between two modifications no solution has been accepted, β is increased by 0.5.

---

[1] We thank C. Cleraux for providing us the code.

As soon as a solution is accepted, β is reset to its value of 0.5. These parameters have been set up according to previous experiments done with the p-median problem [19,21].

4.4   Genetic Algorithm

Several researchers have proposed genetic algorithms (GA) for solving the p-median problem [2,11]. Most of them use a classical string representation, i.e. each chromosome is represented as a single string of length p embedding the index of the selected facilities or nodes. In our experiments, we are using a genetic algorithm based on Correa et al. [11], that has proven its efficiency for large scale instances. We add the constraint that in all chromosomes no facility is duplicated. The initial population of our algorithm is randomly generated and all the candidate locations have the same probability to be chosen. The crossover used is the one described in [11]. It takes as input 2 chromosomes (called A and B) and generates two new offsprings (A' and B') which replace the ones used as input in the global population. The two new chromosomes are generated by the following procedure:

1. Numbers that appear in both A and B are copied into A' and B'
2. Two exchange vectors EA and EB are computed as follows: EA (resp. EB) are integers in A (resp. B) which are not a member of B (resp. A); (obviously EA and EB have the same size).
3. Let r be a random integer between 0 and the size of EA.
4. r integers are randomly selected from EA and EB and exchanged against each other.
5. EA (resp. EB) is copied into A' (resp. B')

We also reuse the mutation of Correa et al. [11]. For a given chromosome, let r be a random integer between 1 and $p/10$. Pick at random r integers from A and for each of them choose at random an integer not in A to replace it. The resulting genetic algorithm will be called basic-GA in the rest of the paper.

Correa et al. [11] also introduce a local search method in their GA by means of a new mutation called *hypermutation*. This operation works as follows for a given chromosome: the heuristic loops on all selected facilities represented by the chromosome. Each selected facility is successively replaced by a facility not already present in the solution, and for every new chromosome produced in this way, the best solution with respect to fitness is kept. hyper-GA is the version of the genetic algorithm which uses hypermutation.

5   Improved Genetic Algorithm

In this section an improved genetic algorithm called imp-GA is proposed. It is designed to perform well on very large p-median problems to be described in Sect. 6. This improved algorithm is based on the hyper-GA of Correa et al. [11]. In hyper-GA, the local search is done via hypermutation and has a complexity of $O(pN)$. Instead, `imp-GA` has a lower complexity for its local search which is shown below as Algorithm 1. It selects a chromosome and returns the best solution found in its neighborhood.

We have designed a new version of the hypermutation operation which turns out to work well on very large problem instances. Its main idea is to diminish the space of the local search in order to reduce the computing cost of hypermutation as soon as we increase the graph size. The new hypermutation has a complexity of $O(N)$ since only a small, fixed number F of selected candidates are considered for replacement.

begin
    Let A be the chromosome;
    Choose randomly F facilities that appear in A;
    foreach *facility among the P chosen nodes* **do**
        for $0 \le i \le N$ **do**
            replace the selected facility by facility i (if i is not already in A);
            if *fitness(A′) < fitness(A)* **then**
                | $A \longleftarrow A'$;
            end
        end
    end
end

Algorithm 1. The new hypermutation operation

In our experiments, we have limited the number of iterations to 100, and the parameter F has been set to 5. Since the new hypermutation has a lower complexity, it is run at each iteration of the genetic algorithm in contrast to that of the hypermutation of the previously described genetic algorithm whose probability of being used for a given generation is only 0.5 %. The new hypermutation has only a reduced local search space, but this leads to small computation times and therefore we can apply it more often. All other parameters have been set to the same values as those used in [11].

We have applied two selections. Correa is using *rank* selection. The *biased roulette wheel* is another selection scheme commonly used [12]. These two selection methods have been tested and experiments have shown that the roulette wheel gives better results than the rank selection. Therefore, for the rest of the article we will only consider the biased roulette wheel selection. It works as follows: we have n chromosomes. Each of them has a fitness value $f_i, i = 1, ..., n$. The probability for an individual i to be chosen is $p(i) = \frac{Max - f_i}{\sum_i Max - f_i}$, where Max is the highest fitness value among all chromosomes of the population. Once the probability of each individual is computed, the selection will choose randomly n new candidates according to their probabilities.

Table 1. Results for the Beasley benchmark (40 graphs)

	CPlex	SA	Genetic algorithms			Volume
			basic-GA	hyper-GA	imp-GA	
			1000 iterations	1000 iterations	100 iterations	
# optimal	40	11	20	20	23	5
Deviation in %	0.0	2.35	0.1	0.14	0.2	4.2
Total time (secs)	64329	2294	70	5246	4862	112

6 Data

For our experiments, we have used two different instance sets. One is the set of graphs from Beasley [7] which is commonly used for testing p-median algorithms. The number of candidate nodes varies from 100 to 900. Before running the tested algorithms, we have precomputed for each instance the matrix of distances.

The second instance set is based on official Swedish data. The distance matrix is derived from the National Road Database (NVDB) of the Swedish Road Administration. From the database which stores all road segments, a graph representing the road network is built by identifying crossings using the x,z,y coordinates [19]. Islands or strongly connected components are detected and virtual links are added to simulate ferries. Then, the graph is cleaned from data unnecessary for the p-median problem such as dead-end roads with no people living along, or points which are neither crossings nor demand nodes. After this process, we still have several millions of nodes. According to Hakimi [16], these points are the set of points we should consider for the p-median location nodes. This graph will be used to compute distances between location and demand nodes. The set of demand nodes is provided by Statistic Sweden [19] and consists of $188, 325$ weighted points. Each point indicates how many persons at the age of 20 to 64 are living in a square centered around this point (in 2012). The size of the square varies from 250 by 250 m to 1 km by 1 km depending on its location. The total size of the population is $5, 411, 373$ persons. The average number of people represented by a point is thus 29 and the highest one $2, 302$.

Observe that the approach used by Carling et al., and Rebreyend et al. [9, 21] for the Swedish province of Dalecarlia cannot be used directly due to the lack of information on the road type. In order to reduce the number of candidate nodes, and to have a good tradeoff between quality and accuracy, we have therefore chosen as candidate node the centers of 1938 Swedish settlements. Some further arguments can be put forward to justify this approach. First, according to previous results of [9, 21], since p will be small (less than 100 for all Sweden), a smaller number of candidate points may not degrade the solution found. Another argument is that all hospitals in Sweden are located less than 3 Km to the city center. Finally, settlements represent well the area where most of the people are located and therefore highly densely populated areas will have more candidate points than sparse areas which is important when the distribution of the demand is non-uniform. Smaller tests were created by restricting locations to the Dalarna province (named Dalarna) and to both the Dalarna and Gävleborg

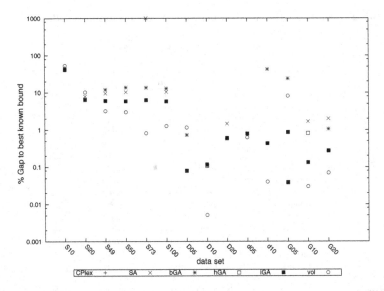

Fig. 2. Compared quality of the solutions obtained by different algorithms on the Swedish benchmark. Quality of a solution is expressed as the percentage of the gap with respect to the best known bound

provinces (named DalGavle). A third smaller case was created by taking only 54 candidate nodes for Dalarna instead of the 108 settlements.

7   Results

We aim to compare the different algorithms described in Sect. 4, with the one we designed, imp-GA, applied to the benchmark set of Beasley [7], and then to the real Swedish data set described in Sect. 6. We want to check the quality of solution and of the runtimes, and also to verify if similar results are obtained for standard benchmark and real-life testcases. For the algorithms' settings, we have used values by default from the corresponding software for CPlex and the Volume algorithm. Since SA is sensitive to its starting point, results are the best of 4 runs from different starting points, each run having 2, 000, 000 iterations.

Table 1 synthesizes our results for the Beasley benchmark set. Average results in terms of quality are perfect for CPlex, solving all of the cases to optimality: the benchmark scale (900 nodes maximum) is affordable by up to date software and hardware. Concerning heuristic approaches, imp-GA solves the most important set of cases to optimality (23) and is always close to, with a standard deviation of 0.2 %. Other GAs also perform well, with similar standard deviations. SA is worst, with a deviation of 2.35 %. The Volume algorithm is the less accurate approach for those cases, with a standard deviation of 4.2 %, but it is the fastest in term of global runtime: 112 s vs 4862 s for imp-GA, and 64329 secs for CPlex. Globally, these results show that all algorithms are applicable to "small" cases solvable to optimality with an exact approach.

Table 2. Results for the Swedish benchmark. (*) excluding Sweden subset

Problem	N	p	Best (known lower bound)	CPLEX	SA	Genetic algorithms basic-GA 1000 iterations	hyper-GA 1000 iterations	imp-GA 100 iterations	Volume
Sweden	1938	10	40553	MEM	58590	59398	TIME	57332	62163
		20	35338	MEM	38154	37720	TIME	37614	38965
		49	19848	MEM	21744	22228	TIME	21042	20487
		50	19633	MEM	21661	22346	TIME	20788	20225
		73	15633	MEM	171961	17766	TIME	16621	15761
		100	12930	MEM	14291	14598	TIME	13689	13095
Dalarna	108	5	19863	19879	19879	20008	19879	19879	20093
		10	11660	11673	11673	11674	11673	11674	11661
		20	7237	7280	7343	7280	7280	7281	7237
Dalarna54	54	5	20910	21075	21075	21075	21075	21075	21041
		10	12270	12323	12323	17510	12323	12323	12275
		20	8398	8472	8472	10404	8472	8472	8398
DalGavle	195	5	27937	27948	27948	27948	27948	27948	30221
		10	17486	17510	17782	17510	17630	17510	17492
		20	10294	10323	10502	10404	10323	10323	10302
% Quality std dev.				0.3(*)	6.7	11.8	0.4(*)	5.0	5.4
Total time (secs)				18128(*)	153595	21717	699(*)	52591	71554

Figure 2 shows our results on the solution quality in a comparative way for our real-case problem instances of large size. The gap between the result of a given method and the best known bound is shown. On the abcisse, Sxx represents the different graphs for the case of Sweden, Gxx represents DalGavle graphs, Dxx Dalarna graphs and dxx dalarna54 graphs. As shown in Table 2, some results are missing, due to unterminated solution processes. The table details the quality results for Sweden data, and also indicates the computational effort via the total runtime of the different algorithms. TIME and MEM indicate that the corresponding algorithm was not able to complete, either within a limited amount of time (2 days) or due to memory limitations (the software aborts before completion on our computer with 32 Gb of memory).

The quality deviation row of Table 2 shows the average deviation in percent between the solution found by the corresponding algorithm and the best lower bound. The best lower bound is the lower bound found either by CPlex or by the Volume algorithm. To compute this standard deviation, we only take into account the set of problem instances for which the algorithm terminates. This explains why algorithms which fail on large graphs (like CPlex) have suprisingly good values. The same explanation holds for global runtimes.

Among the algorithms that can handle all of the Swedish cases, imp-GA provides the best results on average, with a standard deviation of 5.0 %. It performs particularly well for small values of p. Concerning runtimes, as opposed to the Beasley "small" benchmark results, it outperforms the Volume algorithm with a total benchmark runtime of 52591 s vs 71554 s.

Memory utilization is a major concern for the CPlex algorithm which fails on some instances. For the Volume algorithm, almost 22 GB of memory are needed to work on graphs representing Sweden while Simulated Annealing and imp-GA use less than 2.8 GB for the largest cases.

8   Conclusion and Future Work

In this paper, we have compared the quality of several algorithms with respect to real-case instances of the p-median problem, which are large and whose demand is non-uniformly distributed. For this, we have introduced a new benchmark including up to 1938 candidates nodes. The CPlex approach is able to find optimal solutions for all of the Beasley testcases but fails to provide results for our set of very large graphs. Simulated annealing and the basic genetic approach exhibit average results. A hybrid genetic algorithm called hyper-GA has been tested. It improved the results of the basic GA on most of the graphs but failed on the largest graphs. The Volume algorithm has also been tested but its results vary in quality depending on the size of the problem instance.

To obtain better results for the large graphs, we have introduced a new hybrid genetic algorithm called `imp-GA`. This algorithm outperforms all other tested methods on large graphs and has a memory footprint which is as small as that of Simulated Annealing. Its runtime is lower than that of the Volume algorithm. Our results exhibit well the trade-off between exact and approximate methods in dependence on the size of the problem. The effect of a hybrid mutation within a genetic algorithm is important. Therefore, a good design of such a mutation, smartly restricting neighborhood search, leads to an efficient algorithm, especially for large problem instances. Since genetic algorithms can be efficiently parallelized, the proposed method `imp-GA` is a good candidate to deal with large real-case p-median problems.

In the future, we envisage to study other heuristics with respect to large problem instances. Observe, that in our approach distances have been precomputed from the graph. As an alternative we could design methods which extract an interesting set of candidate nodes from a dense graph, or which better exploit the planarity which is typical for graphs arising from geographical problems.

References

1. Al-Khedhairi, A.: Simulated annealing metaheuristic for solving p-median problem. Int. J. Contemp. Math. Sci. **3**(25–28), 1357–1365 (2008)
2. Alp, O., Erkut, E., Drezner, Z.: An efficient genetic algorithm for the p-median problem. Ann. Oper. Res. **122**(1–4), 21–42 (2003)
3. Ashayeri, J., Heuts, R., Tammel, B.: A modified simple heuristic for the p-median problem, with facilities design applications. Robot. Comput. Integr. Manufact. **21**(4–5), 451–464 (2005)
4. Avella, P., Boccia, M., Salerno, S., Vasilyev, I.: An aggregation heuristic for large scale p-median problem. Comput. Oper. Res. **39**(7), 1625–1632 (2012)
5. Avella, P., Sassano, A., Vasil'ev, I.: Computational study of large-scale p-median problems. Math. Program. **109**(1), 89–114 (2007)

6. Barahona, F., Anbil, R.: The volume algorithm: producing primal solutions with a subgradient method. Math. Program. **87**(3), 385–399 (2000)
7. Beasley, J.E.: OR-library: distributing test problems by electronic mail. J. Oper. Res. Soc. **41**(11), 1069–1072 (1990)
8. Cantú-Paz, E.: A survey of parallel genetic algorithms. Calculateurs Parallèles, Réseaux et Systèmes Répartis **10**, 141–171 (1998)
9. Carling, K., Han, M., Håkansson, J.: Does Euclidean distance work well when the p-median model is applied in rural areas? Ann. Oper. Res. **201**(1), 83–97 (2012)
10. Cleraux, C., Bourges, P.: Relaxation Lagrangienne et le problème du p-médian. Master's thesis, Institut Supérieur d'informatique, de modélisation et de leurs applications, Campus de Clermont-Ferrand/Les Cézeaux, BP 10125, 63173 Aubière CEDEX, France (2009)
11. Correa, E., Steiner, M., Freitas, A.A., Carieri, C.: A genetic algorithm for the P-median problem. In: Proceedings of 2001 Genetic and Evolutionary Computation Conference (GECCO-2001), pp. 1268–1275 (2001)
12. Corrêa, R., Ferreira, A., Rebreyend, P.: Scheduling multiprocessor tasks with genetic algorithms. IEEE Trans. Parallel Distrib. Syst. **10**(8), 825–837 (1999)
13. CPlex online reference manual
14. Gay, J.: Résolution du Problème du p-médian, Application à la Restructuration de Bases de Données Semi-Structurées. Ph.D. thesis, Université Blaise-Pascal, Clermont-II (2011)
15. Hakimi, S.L.: Optimum locations of switching centers and the absolute centers and medians of a graph. Oper. Res. **12**(3), 450–459 (1964)
16. Kariv, O., Hakimi, L.: An algorithmic approach to network location problems. SIAM J. Appl. Math. **37**(3), 539–560 (1979)
17. Lim, G.J., Ma, L.: Gpu-based parallel vertex substitution algorithm for the p-median problem. Comput. Ind. Eng. **64**(1), 381–388 (2013)
18. Lim, G.J., Reese, J., Holder, A.: Fast and robust techniques for the Euclidean p-median problem with uniform weights. Comput. Ind. Eng. **57**(3), 896–905 (2009)
19. Meng, X., Rebreyend, P.: From the road network database to a graph for localization purposes. Technical report 2014:09, Dalarna University, Statistics (2014)
20. Mladenoviç, N., Brimberg, J., Hansen, P., Moreno-Pérez, J.A.: The p-median problem: a survey of metaheuristic approaches. Eur. J. Oper. Res. **179**(3), 927–939 (2007)
21. Rebreyend, P., Han, M., Håkansson, J.: How do different algorithms work when applied on the different road networks when optimal location of facilities is searched for in rural areas? In: Huang, Z., Liu, C., He, J., Huang, G. (eds.) WISE Workshops 2013. LNCS, vol. 8182, pp. 284–291. Springer, Heidelberg (2014)
22. Reese, J.: Solution methods for the p-median problem: an annotated bibliography. Networks **48**(3), 125–142 (2006)
23. ReVelle, C.S., Swain, R.W.: Central facilities location. Geogr. Anal. **2**(1), 30–42 (1970)

A New Solution Representation
for the Firefighter Problem

Bin Hu$^{(\boxtimes)}$, Andreas Windbichler, and Günther R. Raidl

Institute of Computer Graphics and Algorithms,
Vienna University of Technology,
Favoritenstraße 9-11/1861, 1040 Vienna, Austria
{hu,raidl}@ads.tuwien.ac.at
windbichler.a@gmail.com

Abstract. The firefighter problem (FFP) is used as a model to simulate
how a fire breaks out and spreads to its surroundings over a discrete time
period. The goal is to deploy a given number of firefighters on strategic
points at each time step to contain the fire in a most efficient way, so that
as many areas are saved from the fire as possible. In this paper we introduce a new solution representation for the FFP which can be applied in
metaheuristic approaches. Compared to the existing approach in the literature, it is more compact in a sense that the solution space is smaller
although the complexity for evaluating a solution remains unchanged.
We use this representation in conjunction with a variable neighborhood
search (VNS) approach to tackle the FFP. To speed up the optimization process, we propose an incremental evaluation technique that omits
unnecessary re-calculations. Computational tests were performed on a
benchmark instance set containing 120 random graphs of different size
and density. Results indicate that our VNS approach is highly competitive with existing state-of-the-art approaches.

Keywords: Firefighter problem · Spreading simulation · Variable
neighborhood search

1 Introduction

The firefighter problem (FFP) was introduced by Hartnell [12] as a model to simulate the spread and containment of fire or disease over a discrete time period in
a simplified way. In each time step the fire (or other malady) infects surrounding
areas whereas a number of firefighters strategically protect certain regions in
order to seal off the respective area and to prevent further spreading. In recent
years this model has also been used to simulate information or virus spreading
in computer networks.

Formally, the problem is defined on an undirected graph $G = \langle V, E \rangle$ where
$|V| = n$ and each vertex $v \in V$ is initially flagged as untouched. At time $t = 1$,

This work is supported by the Austrian Science Fund (FWF) grant P24660-N23.

© Springer International Publishing Switzerland 2015
G. Ochoa and F. Chicano (Eds.): EvoCOP 2015, LNCS 9026, pp. 25–35, 2015.
DOI: 10.1007/978-3-319-16468-7_3

a fire breaks out at a pre-defined set of vertices $B_{init} \subseteq V$, and these vertices are flagged as burnt. For each time step $t = 2, 3, \ldots$, each member of a set of D firefighters protects a vertex $v \in V$ that is not burnt whereas the fire propagates to the neighboring vertices that are untouched and burns them. This process continues until the fire is contained, i.e., it cannot spread any further. The objective is to choose those D vertices in each step to be protected by the firefighters so that when the fire is contained, the number of vertices that are not burnt is maximal. It is assumed that protection is permanent, i.e., a vertex will not catch fire once it is protected, and that firefighters are able to move between arbitrary vertices from one time step to the next one.

An example is shown in Fig. 1. Suppose the number of firefighters is two, i.e., $D = 2$. At time $t = 1$ the fire starts at v_7 (red circle). At $t = 2$ we protect the vertices v_4 and v_9 (blue squares); the fire spreads to vertex v_{10}. At $t = 3$ we protect the vertices v_5 and v_{12}; the fire spreads to vertex v_{11}. At $t = 4$ we protect the vertices v_6 and v_8; the fire is finally contained. The objective value of this solution is 9 saved vertices.

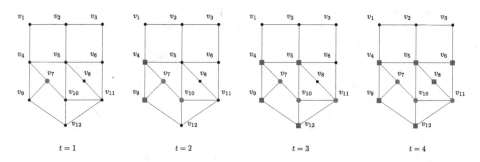

Fig. 1. Example for a graph with 12 vertices and two firefighters (Color figure online).

The contribution of this paper is to introduce a new compact solution representation for metaheuristic approaches for the FFP. We use it in a variable neighborhood search (VNS) approach along with an incremental evaluation technique to boost the performance and compare our results with those existing in the literature.

2   Previous Work

During the last 10 to 20 years, the FFP was studied by several researchers, but mostly from a theoretical point of view. An extensive survey can be found in [8]. The complexity of FFP was studied on various types of graphs: Finbow et al. [9] showed that it is NP-hard on trees with maximum degree three, but solvable in polynomial time if the fire starts at a vertex of degree two. MacGillivray and Wang [17] showed that it is NP-hard on bipartite graphs. King and MacGillivray [16] showed that the FFP is NP-hard if G is a cubic graph. In addition to these graph types, grid structured graphs are particularly interesting due to their relevance in real world scenarios. Properties of the FFP such as the ability to contain the fire was studied by Fogarty [10] and Moeller et al. [19] on two dimensional grids. Later the results were generalized for higher dimensions by Develin and Hartke [7]. Besides the graph structure, complexity also depends on the number of available firefighters. Cases with more than one firefighter were studied by Bazgan et al. [1] and Costa et al. [6]. Apart from the complexity, approximation algorithms for FFP have been extensively studied in the literature, especially the case where G is a tree caught great interest. While Hartnell et al. [13] considered greedy approaches, Hartke [11] proposed a linear programming relaxation based algorithm and Cai et al. [3] proposed a subexponential algorithm. Later the approximation ratios of these approaches were improved by Iwaikawa et al. [15].

Recently Blum et al. [2] presented the so far only existing metaheuristic approach for the FFP based ant colony optimization (ACO). It uses a permutation based solution encoding representing the order of vertices that are considered for protection. The pheromone model contains values for each vertex appearing at each possible position in the permutation. As an extension, the authors also presented a hybrid ACO variant where half of the computation time is spent by the ACO and the latter half is used by further tuning the best solution found by the ACO via mixed integer programming. For this purpose they apply solution polishing which is a mechanism in CPLEX that emphasizes on improving a given solution instead of proving optimality. Surprisingly there are so far no other metaheuristic approaches in the literature to the best of our knowledge. Therefore the (hybrid) ACO is the main competitor for our VNS approach with which we will experimentally compare.

There are other variants of the FFP such as the fractional FFP where firefighters can split their strength into fractions to protect the vertices [10] or the stochastic variant where spreading is non-deterministic [4]. However, those variants are not the scope of this work.

3   Proposed Algorithm

The main aspect of this paper is to introduce a new solution representation for the FFP as a more compact alternative to the permutation based encoding in [2]. We use this new representation in a general variable neighborhood search (VNS) approach with variable neighborhood descent (VND) as embedded local improvement, see [18] for basic information on this kind of metaheuristics.

3.1   Solution Representation

We encode a solution as bitvector $P = \langle p_1, \ldots, p_n \rangle$ where $p_v, v \in V$, is 1 if the vertex should be protected and 0 otherwise. The solution does not explicitly state which vertices to be protected at a particular time step, but this information is implicitly derived during evaluation by Algorithm 1.

Algorithm 1. evaluate(P)

Input: solution $P = \langle p_1, \ldots, p_n \rangle$
Output: repaired solution P and its objective
$\forall v \in B_{\text{init}} : status[v] = -1$;   // the fire starts at time step 1
$\forall v \in V \setminus B_{\text{init}} : status[v] = 0$;   // all other vertices are untouched
$t = 2$;
$freeff = 0$;   // number of available firefighters in each step
$burnt = 0$;   // number of burnt vertices
$burning = true$;
while $burning$ **do**
$\quad$ $burning = false$;
$\quad$ $freeff = freeff + D$;
$\quad$ $neededff = 0$;   // number of currently needed firefighters
$\quad$ **foreach** vertex v adjacent to a burnt vertex at step $t - 1$ **do**
$\quad\quad$ **if** $p_v == 1$ **then**
$\quad\quad\quad$ $neededff = neededff + 1$;

$\quad$ // now each vertex that is adjacent to a just burnt one has to be
$\quad\quad$ updated, i.e., either protected or burnt
$\quad$ **foreach** vertex v adjacent to a burnt vertex at step $t - 1$ **do**
$\quad\quad$ **if** $p_v == 1$ **then**
$\quad\quad\quad$ // if there are too many vertices that should be protected
$\quad\quad\quad\quad$ in the current step, drop a respective number with
$\quad\quad\quad\quad$ uniform probability
$\quad\quad\quad$ **if** $neededff > freeff$ and $\text{random}(1, neededff) > freeff$ **then**
$\quad\quad\quad\quad$ $p_v = 0$;
$\quad\quad\quad\quad$ $neededff = neededff - 1$;
$\quad\quad\quad$ **else**
$\quad\quad\quad\quad$ $status[v] = t$;   // protect vertex at step t
$\quad\quad\quad\quad$ $freeff = freeff - 1$;
$\quad\quad\quad\quad$ $neededff = neededff - 1$;
$\quad\quad$ **if** $p_v == 0$ **then**
$\quad\quad\quad$ $status[v] = -t$;   // burn vertex at step t
$\quad\quad\quad$ $burnt = burnt + 1$;
$\quad\quad\quad$ $burning = true$;
$\quad$ $t = t + 1$;
return $n - burnt$;

During this procedure, we store for each vertex its *status*: 0 for untouched, a positive number z for protected at time step z and a negative number $-z$ for burnt at time step z. The number of burnt vertices is stored in *burnt* and potentially increases over the time steps. The number of available firefighters in each time step is indicated by *freeff* and corresponds to the number of firefighters D minus the number of vertices that would be set on fire at the current step but are marked as protected. This is the actual time when the vertex gets protected. Additionally, spare firefighters in *freeff* can be buffered for future steps in the case not all of them are required at the current time step. Note that the concept of buffering does not conflict with the original problem definition since excess firefighters can be regarded as if they were protecting vertices that would become relevant later, at a time where more than D firefighters are required to protect the desired vertices. Then the number of required firefighters *neededff* is calculated, which consists of the number of vertices that are next to a burning vertex but should be protected according to the solution P. If *neededff* is larger than *freeff* at a certain time, the solution is actually infeasible since there are not enough firefighters to protect all the desired vertices from burning. Therefore, the evaluation procedure also contains an implicit repair function for this situation. If too many vertices are required to be protected, only *freeff* out of *neededff* vertices are saved and the others are dropped on a random basis where each vertex has equal probability. Other than that, the fire spreads over to adjacent vertices that are labeled as untouched.

An example for this procedure is shown in Fig. 2. The number of firefighters is again two and the fire starts at v_7 (red circle) at time $t = 1$. The solution vector is $P = \langle 1, 1, 0, 1, 0, 1, 0, 1, 1, 1, 0, 0 \rangle$ and states that vertices $v_1, v_2, v_4, v_6, v_8, v_9, v_{10}$ should be protected (blue boxes). At $t = 2$ three protected vertices are adjacent to a burning vertex: v_4, v_9 and v_{10}. Since there are only two firefighters, vertex v_4 gets dropped at random and catches fire (red square) while the other two are protected (blue squares). At $t = 3$ only one protected vertex v_1 is adjacent to burning vertices, so one spare firefighter is buffered. The unprotected vertex v_5 catches fire. At $t = 4$ three protected vertices are adjacent to burning vertices and all of them can be saved due to the buffered firefighter. This is equivalent to as if the spare firefighter had saved one of these vertices in the previous time step. Now the fire is fully contained and the objective value of this solution is again 9.

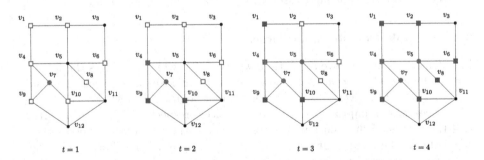

Fig. 2. Solution evaluation example for a graph with 12 vertices and two firefighters (Color figure online).

The whole evaluation procedure that includes the fire-spreading simulation runs in time $O(|V| + |E|)$ since each vertex and each edge is considered only once. This is the same complexity as the evaluation procedure of the ACO in [2].

However, the bitvector representation is more compact as its solution space has a size 2^n instead of $n!$ permutations.

3.2 Initial Solution

We tested two different strategies for creating initial solutions. In variant A, we assign a *closeness centrality* value c_v to each vertex $v \in V \setminus B_{\text{init}}$. This is done by calculating for each vertex the sum of its distances to all other vertices and then taking the inverse value as its closeness. This criterion is often used to determine the influence of a vertex to spread information through a network [20]. A vertex with a small closeness value indicates that once it burns, it has a high potential to further spread the fire. Based on the closeness centrality, we protect each vertex $v \in V \setminus B_{\text{init}}$ with a probability of $\frac{c_{\max} - c_v}{2(c_{\max} - c_{\min})}$ where $c_{\max}$ and $c_{\min}$ are the minimal and maximal closeness values over all vertices $V \setminus B_{\text{init}}$, respectively. This means that the vertex with minimal closeness gets protected with a probability of 50 % whereas the vertex with maximal closeness never gets protected.

In variant B, we just use an empty solution, i.e., where no vertex is set to be protected. This way we entirely rely on the improvement procedures of the VNS to set the protection status.

During preliminary tests, we applied local improvement via VND on these initial solutions and it turned out that no obvious trends between the two variants could be observed. Over a set of 120 instances (see Sect. 4), variant A was better in 32 cases and variant B in 49 cases. However the average solution quality of variant A was better by 0.2 %. On the one hand, it is a good sign that the improvement procedure works so well since it minimizes the differences of initial solutions. On the other hand, we also feel that the construction heuristic certainly has some improvement potentials. Since there was no obvious trend, we used variant B in the later tests since it is the simpler approach.

3.3 Variable Neighborhood Descent

In order to locally improve a solution, we use VND which considers the following neighborhood structure in a best improvement fashion. For neighborhood $N_l(P)$, $l = 1, 2, \ldots$, we consider in the current solution P a set W_l consisting of vertices with l unprotected adjacent vertices, respectively. For each vertex $w \in W_l$, we test if the solution improves by protecting its adjacent vertices, see Algorithm 2. This procedure maximizes the locality of the neighborhood structure since correlated vertices are considered together. If an infeasible solution arises, it is repaired during the evaluation procedure automatically. After that, the *improve* procedure tries to add single vertices to the protection until the solution does not improve.

In order to boost the performance, we implemented an *incremental evaluation scheme* on the evaluation procedure. Whenever we change for a vertex v its protection status p_v, we know the time it was processed in the last evaluation

---

Algorithm 2. VND

Input: solution $P = \langle p_1, \ldots, p_n \rangle$, neighborhood l
Output: Improved solution P
W_l = set of vertices with l unprotected adjacent vertices;
$P_{\text{best}} = P$;
foreach $w \in W_l$ **do**
    $P' = P$;
    $\forall v$ adjacent to $w : p'_v = 1$;
    evaluate(P');
    improve(P');
    if P' better than P_{best} **then**
        $P_{\text{best}} = P'$;

return P_{best};

---

by looking at $status[v]$. Either it was protected at time t if $status[v] > 0$, or it got burnt if $status[v] < 0$. Therefore when we evaluate the new solution after changing p_v in the neighborhood search or in the improve-procedure, we only have to re-calculate vertices $w \in V$ with $|status[w]| > t$ and possibly untouched vertices. In other words, the fire spreading simulation until time t with the status values of corresponding vertices will not change. We only have to take the existing status values from the previous evaluation, calculate the number of spare firefighters *freeff* and set B_{init} to those vertices burnt at time $t - 1$ to continue the fire simulation. If the protection status of more than one vertex gets changed, we have to continue the fire simulation at the smallest time where these vertices were processed. During preliminary tests this incremental evaluation scheme was able to speed up the whole approach by a significant amount. One variant we also considered was to use a first improvement strategy where we could cascade the neighborhood solutions in a way that solutions requiring less re-calculations are evaluated first to maximize the potential gain of the incremental evaluation. However, the best improvement strategy still produced better results in comparable time.

3.4   Variable Neighborhood Search

We use the general VNS framework with shaking when VND is not able to improve the solution any further. A common situation where VND gets stuck in local optima is that when evaluating and repairing an infeasible solution, vertices close to B_{init} are less likely to become dropped from protection than vertices that are farther away. This is due to the repair mechanism that always "activates" the protection for vertices close to B_{init} first. If there are too many vertices marked for protection at a certain time step, it repairs the solution by dropping vertices from the current step. However, dropping vertices from previous time steps would actually also work since excess firefighters would be buffered and carried over to the time where the deficit arises. Since this is not considered in the current implementation, we have a slight bias that prefers vertices close to B_{init}. While this is in fact a reasonable strategy most of the time since protecting vertices close to B_{init} from early on often has a positive impact, there are situations where this behaviour causes the VND to get stuck in local optima.

For this reason, if shaking is called with a size k, it drops the protection status of k vertices selected randomly in the solution so that VND is able to explore new combinations of protected vertices.

4   Computational Results

We tested our VNS on the benchmark set from [2]. It consists of instances with 50, 100, 500, and 1000 vertices with three different edge densities, respectively. The edge density describes the probability of having an edge between two vertices in a random graph and are stated in the Tables 1, 2, 3 and 4. Each combination of vertex count and edge density contains 10 random instances, which results in a benchmark set containing 120 instances in total. In addition, each instance is considered with different number of firefighters $D \in \{1, \ldots, 10\}$. All tests were carried out on a single core of an Intel Xeon E5540 with 2.53 GHz and 3 GB RAM. Compared to the hardware used in [2] for the (hybrid) ACO, it is around twice as fast according to the Standard Performance Evaluation Corporation (SPEC) benchmark[1]. Therefore we use half of the ACO runtime for our VNS which is $n/4$ seconds for each instance to obtain roughly comparable results.

In Tables 1, 2, 3 and 4 we show the results grouped by the number of vertices and the edge density. We compare four different approaches: beside our VNS, the other three are taken from [2] and consist of an exact approach using mixed integer programming solved via CPLEX, the ACO and the hybrid ACO. Each line corresponds to a different number of firefighters D whereas each cell shows the average final solution values over 10 random graphs. Best results are marked bold. In the last two lines of the tables, we give a summary by showing for each

Table 1. Results for graphs with 50 vertices, time limit of 12.5 s for VNS.

D	Edge propability $p_e = 0.1$				Edge propability $p_e = 0.15$				Edge propability $p_e = 0.2$			
	CPLEX	ACO	HyACO	VNS	CPLEX	ACO	HyACO	VNS	CPLEX	ACO	HyACO	VNS
1	**7.4**	**7.4**	**7.4**	7.3	**4.5**	**4.5**	**4.5**	**4.5**	**3.1**	**3.1**	**3.1**	**3.1**
2	**26.6**	26.4	26.5	**26.6**	**9.7**	**9.7**	**9.7**	**9.7**	**7.2**	**7.2**	**7.2**	**7.2**
3	**41.8**	40.9	41.6	41.7	**18.8**	16.5	18.5	18.7	**11.2**	11.1	**11.2**	**11.2**
4	**47.9**	47.8	**47.9**	**47.9**	**31.2**	30.5	30.9	**31.2**	**17.5**	16.0	17.2	17.2
5	**48.5**	**48.5**	**48.5**	**48.5**	**39.1**	36.1	**39.1**	**39.1**	**27.7**	26.1	27.6	27.6
6	**48.8**	**48.8**	**48.8**	**48.8**	**43.7**	42.7	**43.7**	**43.7**	**33.0**	31.4	**33.0**	32.9
7	**49.0**	**49.0**	**49.0**	**49.0**	**46.3**	45.4	**46.3**	**46.3**	**37.5**	35.7	**37.5**	**37.5**
8	**49.0**	**49.0**	**49.0**	**49.0**	**48.1**	46.8	**48.1**	**48.1**	**42.7**	40.3	42.6	42.6
9	**49.0**	**49.0**	**49.0**	**49.0**	**48.6**	48.2	**48.6**	**48.6**	**46.1**	44.4	**46.1**	**46.1**
10	**49.0**	**49.0**	**49.0**	**49.0**	**48.8**	**48.8**	**48.8**	**48.8**	**47.5**	47.1	**47.5**	**47.5**
$\sum$	**417.0**	415.8	416.7	416.8	**338.8**	329.2	338.2	338.7	**273.5**	262.4	273.0	272.9
%	**100.00%**	99.71%	99.93%	99.95%	**100.00%**	97.17%	99.82%	99.97%	**100.00%**	95.94%	99.82%	99.78%

---

[1] www.spec.org/cpu2006.

Table 2. Results for graphs with 100 vertices, time limit of 25 s for VNS.

D	\multicolumn{4}{Edge propability $p_e = 0.05$}				\multicolumn{4}{Edge propability $p_e = 0.075$}				\multicolumn{4}{Edge propability $p_e = 0.1$}			
	CPLEX	ACO	HyACO	VNS	CPLEX	ACO	HyACO	VNS	CPLEX	ACO	HyACO	VNS
1	9.2	9.1	9.2	9.1	5.4	5.4	5.4	5.4	3.9	3.9	3.9	3.9
2	26.9	25.7	27.6	27.7	11.3	11.2	11.3	11.2	8.5	8.3	8.7	8.7
3	62.8	54.6	62.7	63.6	41.5	41.0	41.6	41.5	21.4	21.0	21.3	21.4
4	85.3	66.3	85.5	86.0	53.7	52.4	53.3	53.8	25.5	24.5	25.5	25.7
5	97.3	92.3	97.3	97.3	65.7	63.5	65.9	66.5	30.2	29.1	29.5	30.1
6	98.5	98.3	98.5	98.5	87.5	75.1	87.3	87.6	41.8	33.9	41.0	42.3
7	98.8	98.8	98.8	98.8	98.1	87.9	98.1	98.1	58.7	46.4	56.3	56.9
8	98.9	98.9	98.9	98.9	98.6	93.5	98.6	98.6	74.8	62.0	74.0	74.8
9	99.0	99.0	99.0	99.0	98.8	98.8	98.8	98.8	89.2	77.3	88.0	89.2
10	99.0	99.0	99.0	99.0	99.0	99.0	99.0	99.0	94.7	85.9	94.4	94.6
$\sum$	775.7	742.0	776.5	777.9	659.6	627.8	659.3	660.5	448.7	392.3	442.6	447.6
%	99.70%	95.37%	99.81%	99.99%	99.83%	95.02%	99.79%	99.97%	99.80%	87.26%	98.44%	99.56%

Table 3. Results for graphs with 500 vertices, time limit of 125 s for VNS.

D	\multicolumn{4}{Edge propability $p_e = 0.015$}				\multicolumn{4}{Edge propability $p_e = 0.02$}				\multicolumn{4}{Edge propability $p_e = 0.025$}			
	CPLEX	ACO	HyACO	VNS	CPLEX	ACO	HyACO	VNS	CPLEX	ACO	HyACO	VNS
1	7.6	7.5	7.8	7.6	5.3	5.2	5.6	5.7	4.2	4.4	4.3	4.5
2	5.6	13.0	13.6	14.6	10.6	10.4	11.2	11.3	9.1	8.5	9.0	9.3
3	3.1	18.8	21.3	22.3	60.3	63.1	63.6	65.0	12.8	12.7	13.8	13.7
4	150.2	119.9	168.3	170.3	69.5	67.6	70.4	70.0	17.7	16.9	18.6	18.1
5	250.5	218.9	265.9	267.5	45.6	72.7	74.6	75.1	6.5	21.6	22.4	23.2
6	349.1	268.8	363.0	362.9	102.4	123.8	126.5	126.1	25.6	26.3	33.8	28.7
7	448.9	407.6	453.5	453.7	102.7	128.1	130.1	133.1	78.1	77.6	93.0	80.1
8	449.1	453.9	455.0	454.7	299.0	135.8	315.6	316.1	154.2	127.6	173.5	175.5
9	449.1	454.7	456.6	456.9	349.0	317.5	363.8	364.1	223.2	221.8	225.4	223.8
10	498.8	455.5	498.8	498.8	409.2	321.1	410.2	409.3	221.5	225.1	229.6	227.3
$\sum$	2612.0	2418.6	2703.8	2709.3	1462.6	1245.3	1571.6	1575.8	753.0	742.4	823.5	804.2
%	96.39%	89.25%	99.77%	99.98%	92.72%	78.94%	99.63%	99.89%	91.09%	89.80%	99.61%	97.28%

Table 4. Results for graphs with 1000 vertices, time limit of 250 s for VNS.

D	\multicolumn{4}{Edge propability $p_e = 0.0075$}				\multicolumn{4}{Edge propability $p_e = 0.01$}				\multicolumn{4}{Edge propability $p_e = 0.125$}			
	CPLEX	ACO	HyACO	VNS	CPLEX	ACO	HyACO	VNS	CPLEX	ACO	HyACO	VNS
1	105.2	107.4	107.8	108.0	4.9	5.7	6.0	6.3	4.0	4.9	4.7	5.3
2	107.9	112.7	115.0	114.6	4.4	10.9	10.6	12.1	8.0	9.4	10.1	10.2
3	101.7	118.0	118.0	122.1	13.7	15.9	17.0	17.9	4.6	14.3	13.9	15.4
4	399.4	318.3	415.7	417.9	14.8	21.4	24.1	24.1	99.9	116.6	116.5	117.8
5	399.6	419.0	421.2	422.8	104.3	27.1	123.6	126.6	103.3	120.8	120.9	122.8
6	598.8	423.6	614.2	617.1	201.5	129.1	226.0	228.2	99.9	125.5	126.2	128.7
7	898.1	523.5	902.5	903.3	299.7	525.5	325.2	328.2	199.8	226.7	226.6	228.8
8	998.2	528.6	998.2	998.2	399.5	329.7	424.4	426.5	218.7	231.1	234.8	233.9
9	998.9	905.6	998.9	998.9	399.6	427.9	427.7	431.8	301.0	236.2	332.1	332.6
10	999.0	999.0	999.0	999.0	499.5	432.6	528.4	530.7	602.2	335.0	620.5	619.6
$\sum$	5606.8	4455.7	5690.5	5701.9	1941.8	1725.8	2113.0	2132.4	1641.4	1420.8	1806.3	1815.1
%	98.33%	78.14%	99.79%	99.99%	91.06%	80.93%	99.09%	100.00%	90.34%	78.20%	99.42%	99.90%

approach the summed up average solution values ($\sum$) and the percentage of the best values reached (%) when considering for each line the best performing approach.

As reported in [2], CPLEX was able to consistently solve all instances with 50 vertices to optimality. On larger instances, it was usually terminated after reaching the time limit. Especially instances with dense graph and a low number of firefighters proved to be difficult. In these cases CPLEX is outperformed by the metaheuristic approaches. We observe that our VNS performs slightly better than the hybrid ACO. On the majority of the instance sets it is better and in some cases it is worse by a small margin. Compared to the pure ACO approach the VNS is consistently better. Overall, it seems that VNS performs better on sparse graphs and on larger instances. The latter is due to the reduced search complexity of the bitvector representation. On sparse graphs we suspect that it is more convenient for the VND neighborhood structure to iterate through the vertices since their degrees are lower and more diverse. Looking at the closeness of the different approaches, we also think that being able to solve a considerable part of the instances in this benchmark set optimally by CPLEX shows that they are not very hard, thus the margin for improvement is rather slim. Therefore more sophisticated approaches in the future should be tested on more complex instances so that differences become more obvious.

5   Conclusions and Future Work

We proposed a variable neighborhood search approach for the firefighter problem based on a bitvector solution representation. By storing for each vertex only its protection status, it is more compact than a permutation based representation. We also proposed an incremental evaluation technique to speed up the computation significantly. Although the VNS is not able to outperform the hybrid ACO approach in a substantial way, it is typically at least as good when it comes to solution quality and performs significantly better than the standard ACO.

In future work we want to investigate approaches that make use of both representations and associate neighborhood structures since they can be considered as complementary to each other: The bitvector representation stores the protection status but the actual order in which the vertices are protected is obtained during evaluation. The permutation representation stores the order in which the vertices are considered for protection but whether a vertex is protected or not is determined during evaluation. It has been shown that solution methods on combinatorial optimization problems can substantially benefit from using complementary representations, e.g., in the case of generalized minimum spanning tree problem as shown in [5,14]. Considering the FFP, using both representations either in a VNS fashion or infusing the ACO from [2] with a local search method based on the new representation appears to be particularly promising.

References

1. Bazgan, C., Chopin, M., Ries, B.: The firefighter problem with more than one firefighter on trees. Discrete Appl. Math. **161**(7–8), 899–908 (2013)
2. Blum, C., Blesa, M.J., García-Martínez, C., Rodríguez, F.J., Lozano, M.: The firefighter problem: application of hybrid ant colony optimization algorithms. In: Blum, C., Ochoa, G. (eds.) EvoCOP 2014. LNCS, vol. 8600, pp. 218–229. Springer, Heidelberg (2014)

3. Cai, L., Verbin, E., Yang, L.: Firefighting on trees: $(1 - 1/e)$–approximation, fixed parameter tractability and a subexponential algorithm. In: Hong, S.-H., Nagamochi, H., Fukunaga, T. (eds.) ISAAC 2008. LNCS, vol. 5369, pp. 258–269. Springer, Heidelberg (2008)
4. Comellas, F., Mitjana, M., Peters, J.G.: Broadcasting in small-world communication networks. In: Proceedings of the 9th International Colloquium on Structural Information and Communication Complexity, pp. 73–85 (2002)
5. Corus, D., Lehre, P.K., Neumann, F.: The generalized minimum spanning tree problem: A parameterized complexity analysis of bi-level optimisation. In: Proceedings of the 15th Annual Conference on Genetic and Evolutionary Computation (GECCO), pp. 519–526. ACM (2013)
6. Costa, V., Dantas, S., Dourado, M., Penso, L., Rautenbach, D.: More fires and more fighters. Discrete Appl. Math. **161**(16–17), 2410–2419 (2013)
7. Develin, M., Hartke, S.: Fire containment in grids of dimension three and higher. Discrete Appl. Math. **155**(17), 2257–2268 (2007)
8. Finbow, S., Science, C., Scotia, N., MacGillivray, G.: The firefighter problem: a survey of results directions and questions. Aust. J. Comb. **43**, 57–77 (2009)
9. Finbow, S., King, A., MacGillivray, G., Rizzi, R.: The firefighter problem for graphs of maximum degree three. Discrete Math. **307**(16), 2094–2105 (2007)
10. Fogarty, P.: Catching the fire on grids. Master's thesis, University of Vermont, USA (2003)
11. Hartke, S.: Attempting to narrow the integrality gap for the firefighter problem on trees. In: DIMACS Series in Discrete Mathematics and Theoretical Computer Science, pp. 225–231 (2006)
12. Hartnell, B.: Firefighter! An application of domination. In: 20th Conference on Numerical Mathematics and Computing, pp. 218–229 (1995)
13. Hartnell, B., Li, Q.: Firefighting on trees: How bad is the greedy algorithm? In: Proceedings of the Thirty-first Southeastern International Conference on Combinatorics, Graph Theory and Computing, pp. 187–192 (2000)
14. Hu, B., Leitner, M., Raidl, G.R.: Combining variable neighborhood search with integer linear programming for the generalized minimum spanning tree problem. J. Heuristics **14**(5), 473–499 (2008)
15. Iwaikawa, Y., Kamiyama, N., Matsui, T.: Improved approximation algorithms for firefighter problem on trees. IEICE Trans. **E94.D**(2), 196–199 (2011)
16. King, A., MacGillivray, G.: The firefighter problem for cubic graphs. Discrete Math. **310**(3), 614–621 (2010)
17. MacGillivray, G., Wang, P.: On the firefighter problem. J. Comb. Math. Comb. Comput. **47**, 83–96 (2003)
18. Mladenović, N., Hansen, P.: Variable neighborhood search. Comput. Oper. Res. **24**(11), 1097–1100 (1997)
19. Moeller, S., Wang, P.: Fire control on graphs. J. Comb. Math. Comb. Comput. **41**, 19–34 (2002)
20. Newman, M.J.: A measure of betweenness centrality based on random walks. Soc. Netw. **27**(1), 39–54 (2005)

A Variable Neighborhood Search Approach for the Interdependent Lock Scheduling Problem

Matthias Prandtstetter[(✉)], Ulrike Ritzinger, Peter Schmidt,
and Mario Ruthmair

AIT Austrian Institute of Technology, Mobility Department – Dynamic
Transportation Systems, Giefinggasse 2, 1210 Vienna, Austria
{matthias.prandtstetter,ulrike.ritzinger,mario.ruthmair}@ait.ac.at

Abstract. We investigate a so far not examined problem called the *Interdependent Lock Scheduling Problem*. A *Variable Neighborhood Search* approach is proposed for finding lock schedules along the Austrian part of the Danube River in order to minimize the overall ship travel times. In computational experiments the performance of our approach is assessed and compared to real-world ship trajectories. Notable improvements can be achieved. In addition, the number of (empty) lockages can be significantly reduced when taking them into account during optimization without loosing too much of quality in travel time optimization.

Keywords: Interdependent lock scheduling problem · Variable neighborhood search

1 Introduction

Inland navigation can be seen as the most efficient means of transport with respect to ecological objectives [3]. In combination with transportation via trucks and trains it builds a strong overland transportation network. Considering today's transportation volume and emissions related to (overland) transport in Europe [7] a shift of transports from trucks towards trains and especially inland navigation within the next years is highly desired by the European Commission [6].

Among European inland waterways Rhine and Danube are two of the largest (and most important) ones. While the Rhine heads North-South, mainly connecting Switzerland, France, the West of Austria, Germany, and the Netherlands, to the North Sea, the Danube heads West-East, connecting the South of Germany, Austria, Slovakia, Hungary, Croatia, Serbia, Bulgaria, Romania, Moldova, and the Ukraine, to the Black Sea. In contrast to the Rhine, ships traveling on the (Austrian part of the) Danube (approx. 350 m) have to pass nine watergates at

This work is partially funded by the Austrian Federal Ministry for Transport, Innovation and Technology (BMVIT) within the strategic programme I2VSplus under grant 835771 (imFluss).

G. Ochoa and F. Chicano (Eds.): EvoCOP 2015, LNCS 9026, pp. 36–47, 2015.
DOI: 10.1007/978-3-319-16468-7_4

power plants used for electricity production. Naturally, these watergates build a bottleneck in the transportation network as ships have to pass through them to overcome the height difference caused by retaining the river.

It can be seen in the action plan of the Danube Region Strategy [4,5] that only 10 %–20 % of the potential volume is transported on the Danube. Therefore, a significant increase in the transport volume will lead to congestion around watergates [14].

In the Austrian funded project *imFluss*, the main goal is to revise the currently applied *first-come, first-serve* strategy for the scheduling process at watergates and investigate the increase of efficiency by applying alternative scheduling strategies. The slot management is based on an optimization approach which schedules ships at watergates such that the overall sum of travel times of the ships is minimized. Furthermore, it allows to reduce congestion at watergates by providing travel speed advises to the captains. Since ships must traverse multiple watergates in short distances at the Danube, the interdependence of the schedules at the watergates has to be considered as well.

Within this paper, we first give a detailed description of the problem domain and then present a Variable Neighborhood Search (VNS) framework which incorporates various Local Search operators. Computational results show the positive impact of the proposed method on the planning approach, and finally conclusions sum the work up.

2   Related Work

The *Interdependent Lock Scheduling Problem* (ILSP), as addressed here, is not yet defined in the (scientific) literature to the best of our knowledge. However, various related problem definitions and corresponding solution approaches can be found in the literature where the most important ones are outlined in this section. The simplest version of the *lock scheduling problem* (LSP) optimizes the traffic flow through a single watergate which consists of a single lock chamber. The objective is to find a schedule involving up- and downstream vessels which maximizes the traffic flow. In [2], a polynomial-time dynamic program is proposed to solve this LSP. An extended version of this basic problem is proposed by Verstichel and Vanden Berghe [13], where multiple parallel lock chambers with different sizes are available. In addition, they integrate a kind of packing problem which focuses on the (optimized) placement of ships inside the lock chamber. They minimize the waiting times of the ships at watergates as the primary objective, but also consider the minimization of the number of lockage operations as a secondary goal. Beside heuristics for creating initial solutions, metaheuristic approaches are applied for improving the so far found solutions. These metaheuristics include a variable neighborhood search (VNS), a multiple neighborhood search and a composite neighborhood search approach. In [12], Verstichel proposes an exact approach based on integer linear programming. However, this method has unpredictable (long) runtimes which led to a low acceptance by lock masters. While all of these approaches focus on single watergates only, there are

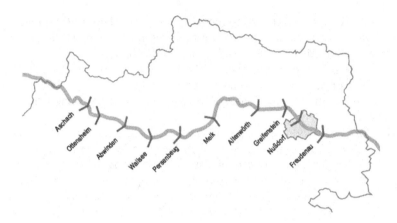

Fig. 1. Sketch of the Austrian part of the Danube with marks where watergates are located. Parts of the river between two watergates (or the state border) are referenced as sections. ©viadonau and DoRIS (http://www.doris.bmvit.at)

some works which consider the ILSP at the Upper Mississippi River in USA [9]. However, the focus in this work is not combinatorial optimization but an estimation of delays based on traffic volume and interdependence among watergates. In [10,11], the authors investigate various strategies which are applied in vessel scheduling and report that a *shortest processing time first* strategy with fairness constraints is more efficient than the classical *first come, first serve* strategy.

3   Problem Description

In the *Interdependent Lock Scheduling Problem* (ILSP) we are given a set of m watergates $\mathcal{G} = \{1, \dots, m\}$, successively arranged along the river. The river is divided into sections which are bounded by watergates or national borders, as it is depicted in Fig. 1 for the Austrian part of the Danube. Let $\mathcal{S}$ be the set of ships which is partitioned into ships $\mathcal{S}^+$ going upstream and ships $\mathcal{S}^-$ going downstream, i.e., $\mathcal{S} = \mathcal{S}^+ \cup \mathcal{S}^-$, $\mathcal{S}^+ \cap \mathcal{S}^- = \emptyset$. Each ship $s \in \mathcal{S}$ has to pass through one or more successive watergates $\mathcal{G}_s \subseteq \mathcal{G}$, either up- or downstream.

Each gate $g \in \mathcal{G}$ consists of two identical and asynchronously operable lock chambers which allow the ships to overcome the water level difference when passing through a watergate. The length of a lock chamber is denoted by κ_g. A lockage operation is defined to either fill or empty the lock chamber with water. The time required for this operation is given by τ_g. Within a lock chamber ships are positioned in a row possibly allowing multiple ships to pass through a gate in a single lockage operation. The ships' positions in a lock chamber define their entering and exiting sequence.

The trip of ship $s \in \mathcal{S}$ starts at a given time and ends when it arrives at its target position. We assume that there are no planned stops for a ship. Additionally, information about the ship length k_s, the ship type r_s, and the average travel speed v_s is available. According to v_s and the positions of the watergates and borders, the travel time for each ship on each river section can be determined. Whenever a ship arrives at a watergate and takes part in a lockage operation it requires some time to enter and to exit the lock chamber, respectively.

In a feasible solution for the ILSP, each ship $s \in \mathcal{S}$ must be assigned to exactly one lock chamber position in exactly one lockage operation at each gate $g \in \mathcal{G}_s$, while respecting the following constraints:

- It must be ensured that only upstream going ships $s \in \mathcal{S}^+$ can be assigned to a lockage operation in which the lock chamber is filled up with water, whereas only downstream going ships $s' \in \mathcal{S}^-$ can be allocated to a lockage operation emptying a chamber.
- The sum of lengths of all ships assigned to the same lockage operation must not exceed the length of the lock chamber κ_g.

Additionally, for each lockage operation a starting time has to be set, based on the following limitations:

- A simultaneous operation of both lock chambers at a watergate is not possible. Therefore, previous lockage operations and the corresponding ships' exiting process have to be finished before the next ships can start their entering process. Only after all ships assigned to the same lockage operation have finished their entering process, the lockage operation is allowed to start.
- Ships are only allowed to enter (or exit) the lock chamber one after another because of safety and space limitations at the gates.
- All ships have to arrive at their target position before planning horizon H.

Because of these constraints, waiting times may arise for some of the ships. The optimization goal of the ILSP is to find a feasible solution such that the sum of total travel times of all ships is minimized, implicitly minimizing the sum of waiting times over all ships. As a second weighted term in the objective function, we optionally consider the minimization of the number of lockage operations since each lockage reduces the amount of water to be used in electricity production.

4 Solution Representation and Decoding

As mentioned in the previous section, a solution in the ILSP is defined by the lockage starting times and a unique assignment of each ship $s \in \mathcal{S}$ to a lock chamber position in a lockage operation for all watergates $\mathcal{G}_s$. Thus, a solution can be stored as a three-dimensional array where the first index refers to a watergate, the second one to a lockage operation at the watergate and the third one to the position of the ship in the lock chamber. The actual values stored in this three-dimensional array correspond to the ship ids. This array can be seen as a *relative schedule* without actual lockage starting times.

Algorithm 1. FIRSTFITCONSTRUCTION

---

1 $Q \leftarrow$ chronologically ordered queue of watergate passing events
2 **while** $Q \neq \emptyset$ **do**
3      $e \leftarrow Q.pop()$
4      $g \leftarrow watergate(e)$
5   **if** *e can be assigned to the currently last scheduled lockage at g* **then**
6 | assign e to last lockage at g
7   **else**
8 | append new lockage(s) at g
9 | assign e to now last lockage
10   compute $\forall e' \in Q$ dependent on e new earliest possible lockage times
11   chronologically re-order Q

Based on this representation and the instance data we propose a decoding procedure to compute the corresponding optimal lockage starting times with respect to minimizing the sum of waiting times. Note that the number of lockage operations is implicitly given in the array. It can be easily seen that choosing the earliest possible starting time for each lockage results in a minimal sum of ship travel times with respect to the current solution. The earliest possible time for lockage l at watergate g depends on

- the previous lockage at gate g (if it exists),
- all lockages at neighboring gates which have at least one ship assigned which is also assigned to l and which has to pass the neighboring gate directly before g, and
- the start times of ships which start their trips in a neighboring section and are assigned to l.

Based on a *relative schedule*, a directed dependency graph can be built with nodes representing lockages and arcs describing a dependency of the lockage at the head on the lockage at the tail of the arc. In case of feasible *relative schedules* this graph has to be acyclic. Then, the starting times for all the lockage operations can be computed by traversing the dependency graph such that a node is only processed (i.e., a lockage starting time is set) if all preceding nodes have already been completed (i.e., starting times of previous lockages at the same or neighboring watergates have been fixed).

5   First Fit Construction Heuristic

In order to find an initial solution which can be further improved by the *Variable Neighborhood Search* framework described in the next section, we present a *First Fit* construction heuristic. It basically iterates over all watergate passages of ships in a chronological order and assigns them to lockages, see Algorithm 1 for a pseudo-code of the approach.

In line 5 it is decided whether a vessel can be added to the (currently) last lockage operation at the corresponding watergate. This decision includes structural decisions based on the length of the lock chamber as well as the traveling direction of the ship. In addition, operational decisions are included such as the maximum shift in time caused by inclusion of a vessel to a lockage. In our implementation, additional lockages are added if a lockage will be postponed more than 30 min, which reflects the amount of time needed for filling (or emptying) a lock chamber.

6   Variable Neighborhood Search Framework

To improve initial solutions we employ a *Variable Neighborhood Search* (VNS) framework [8]. As a local search method within VNS we incorporate *Variable Neighborhood Descent* (VND) [8]. The main idea of VNS and VND is to systematically examine different neighborhood structures such that local optima with respect to single neighborhood structures can be overcome.

6.1   Neighborhood Structures for VND

The neighborhood structures incorporated into our VND implementation are based on the definition of so-called move operators which are used for (slightly) modifying a given solution by small local adjustments. As it is possible to decode a given *relative schedule* such that lockage starting times are optimal with respect to the sum of travel times, no neighborhood structures are defined on varying lockage starting times, cf. Sect. 4. All defined moves work on the *relative schedule* only. In the following we describe the used neighborhood structures.

Shift Vessels. The main idea of this neighborhood structure is to move one or more vessels from one lockage to another where the number of ships to be shifted is an input parameter. For implementation issues we decided to allow only multiple ships to be moved together if they are assigned to the same lockage. In general, a shift can only be performed if

- the corresponding watergate stays the same,
- no circular dependencies are introduced in the underlying dependency graph (cf. Sect. 4), and
- the ships to be shifted fit into the target lock chamber together with all currently assigned ships.

However, it is not only necessary to decide to which other lockage vessels are shifted but also at which actual position the ships will be inserted, i.e., the ordering of the ships in the target lock chamber has to be considered. Note, however, that the relative order of the shifted ships and the ships in the target lock chamber do not change.

Based on this definition, the size of this neighborhood can be estimated by a function linear in the number of lockages per watergate times the maximum number of vessels concurrently assigned to a lockage.

Swap Vessels. As indicated by the name of this neighborhood structure, the main idea here is to swap the positions and/or lockages of two vessels. Basically, the same precondition applies as for shift moves. However, a swap of vessels scheduled for the same lockage is explicitly included in this move type.

The size of a corresponding neighborhood can therefore be estimated to be quadratic in the number of ships passing a watergate. This implies that on average neighborhoods based on this structure will be significantly larger than neighborhoods based on shifting operations.

Remove Empty Lockages. Since a Shift Vessel move may result in empty lockages, it is beneficial to remove them with respect to minimizing the number of lockages. However, removing a single lockage is in most cases not feasible as each lock chamber has alternating upstream and downstream lockages. Thus, we try to remove two lockages at the same time. Therefore, the size of the neighborhood can be estimated by a function which is quadratic in the number of empty lockages.

6.2 Neighborhood Structures for VNS

Analogously to VND, several neighborhood structures are defined to be used within VNS for shaking operations. To keep things simple and fast, we decided to perform a pre-defined number of random shift moves during the shaking phase. Currently, four neighborhood structures are defined with neighborhood structure i performing i random shifts, with $i \in \{1, 2, 3, 4\}$.

7 Experiments and Computational Results

In order to test the performance of the proposed VNS framework and the applied neighborhood structures we conducted a set of computational experiments. For this purpose, we decided to generate a set of test instances which were extracted from real-world data provided by our project partner viadonau who is providing the Austrian River Information Services called DoRIS. Via DoRIS we got anonymized sample data including the trajectories of all vessels for selected days in the period from August 2013 until April 2014. We selected uniformly 30 days in this period. Since the selected period covers days with both low and high traffic demand a broad range of situations and especially number of trips (from 47 up to 151, cf. Table 1) has been investigated. Note that there are days with even more trips along the Austrian Danube. We removed, however, all trips which did not pass at least one watergate as those trips have no particular influence on the ILSP. A real-world scenario was simulated by using the start and target positions as well as departure times of ships as input. Travel times were estimated via the method proposed in [1]—an approach which turned out to be highly accurate. Finally, we compared historic travel times with the output of our algorithm. All of our tests were performed on a single core of a Intel Xeon 2600 processor with 4 GB memory per core (although the average RAM consumption was much

Table 1. Results when only minimizing the total travel time (VNS0) and when additionally minimizing the number of lockages with weight 1000 (VNS1000). Column *hist vs. VNS0* depicts the average relative performance in percent of VNS0 with respect to historical data. Columns *VNS0 vs. VNS1000* show the changes in the relevant key performance indicators (total travel time, number of empty lockages, algorithm runtime) for VNS1000 compared to VNS0.

Instance	#trips	hist. vs. VNS0	VNS0 vs. VNS1000		
		travel time [%]	travel time [%]	#lockages [%]	#empty [%]	runtime [%]
123_2014-01-07	56	−13.50	−0.01	+0.00	+0.00	+3.09
145_2014-02-03	65	−18.21	+0.05	+0.00	−0.44	+41.42
125_2014-01-12	67	−12.68	+0.00	−1.14	−3.75	+5.78
194_2014-02-04	69	−16.21	−0.00	−0.09	−0.62	+12.11
203_2014-03-06	71	−12.70	−0.08	−1.29	−4.81	+11.80
167_2013-12-08	78	−9.38	+0.01	−0.96	−4.00	−11.08
179_2013-12-02	82	−14.02	+0.19	+0.00	+0.05	+15.44
182_2014-01-10	83	−13.49	−0.03	−2.21	−9.55	+2.28
221_2013-11-07	84	−13.01	−0.13	−5.87	−20.54	+11.42
203_2013-09-05	87	−8.73	+0.04	−2.77	−12.63	+32.44
197_2014-03-07	90	−5.87	+0.07	−2.27	−8.23	+1.50
188_2014-02-07	94	−5.55	+0.18	−1.51	−11.32	−17.47
180_2013-12-07	98	−9.82	+0.07	+0.11	+0.26	−9.80
208_2014-01-09	99	−7.45	+0.06	−5.07	−20.29	+11.59
254_2013-10-07	102	−4.17	+0.02	−1.81	−6.61	+14.94
184_2013-11-09	104	−8.50	−0.02	−3.19	−13.26	−1.03
118_2013-08-12	107	−5.62	+0.09	−2.82	−11.73	−9.56
234_2013-09-02	108	−5.82	−0.09	−4.14	−15.16	+11.74
210_2013-10-13	111	−0.08	+0.03	−3.04	−12.30	+12.01
258_2013-12-06	115	−11.58	+0.01	−2.19	−13.08	−9.59
134_2013-08-05	119	−12.14	−0.18	−5.55	−26.40	+27.01
138_2013-08-06	122	−3.33	−0.09	−2.24	−7.52	+12.00
250_2013-10-12	124	−0.81	+0.12	−3.33	−11.60	+21.73
268_2013-10-09	127	+1.89	−0.05	−2.53	−10.26	+7.07
299_2013-09-04	128	−6.19	+0.21	−1.11	−4.50	−13.93
289_2013-12-05	130	−5.10	+0.18	−0.40	−5.16	+42.45
156_2013-08-08	138	−9.26	+0.32	−7.64	−34.21	−13.32
283_2013-10-11	141	+7.56	−0.31	−4.93	−20.05	+18.94
332_2013-09-07	149	+0.47	−0.16	−1.63	−5.82	−6.84
358_2013-09-06	156	+8.52	+0.36	−3.00	−14.56	−9.31

below that threshold). In addition, 30 runs were performed for each test instance and algorithmic setup.

The actual ordering of neighborhood structures in the VND part of the optimization process was chosen such that first a shift of one vessel is examined, followed by a swap of two ships, a shift of two vessels and a shift of three vessels. Finally, we apply the Remove Empty Lockages neighborhood search. A next improvement step function was employed for all neighborhood structures.

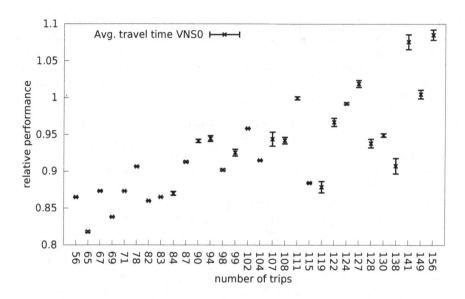

Fig. 2. Average total travel times (with standard deviations) over 30 runs of VNS0 compared to historical data.

For the historic data provided via DoRIS we currently only have information of the ship trajectories but not on the actual lock operations. Therefore, we decided to compare in a first step the measured total travel time of all ships, with the estimated total travel time when applying a lock schedule as proposed by the VNS framework. As the number of lockages for the historic data is not yet known, we conducted two independent test series where in the first, the number of lockages was not taken into account (in the further context referred to as VNS0) while for the second one additional lockage is worth 1000 extra seconds travel time, i.e., we have chosen a weighting ration of 1:1000 between travel time and number of lockages (referred to as VNS1000 in the following). In Table 1 the obtained results can be seen in detail.

A graphical representation is shown in Figs. 2 and 3. While in Fig. 2 the relative improvement of the VNS setups with/without respecting the number of lockages is given, Fig. 3 shows the relative performance of VNS1000 with respect to VNS0. The areas around the averages represent standard deviations.

An improvement with respect to the historic trajectories can be found in almost all cases. However, for some instances the historic solution could not be beaten. This can be explained by two simple reasons: first, although the method for estimating travel times is highly accurate, it might still happen that travel times are underestimated. Second, the estimation how many ships can be packed into one lock chamber is highly heuristic and there are situations where the lockmaster was highly efficient while our heuristic approach decided that putting two specific ships at the same time in the lock chamber is not feasible. However, more important is the observation that VNS1000

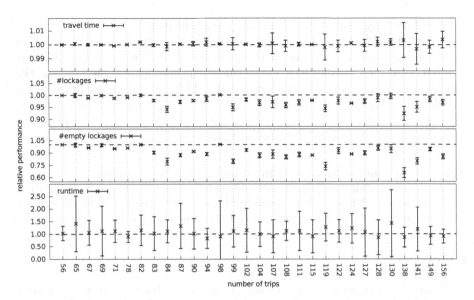

Fig. 3. Relative comparison between VNS0 and VNS1000. The baseline (dashed line) represents the performance of VNS0. The average performance of VNS1000 in relation to VNS0 is plotted together with the standard deviations.

results—although having a more complex objective function—in slightly better results with respect to travel times.

In addition, it can be seen that VNS1000 results in noticeable less empty lockages which is highly important: Although the objective is to maximize the traffic flow (i.e. to minimize the travel times) the additional objective of minimizing the number of lockages is justified by the fact that the embankment dams (in Austria) were erected for energy producing reasons. While lockages with ships can be argued towards an energy supplier, empty lockages are harder to be explained. Therefore, reducing the number of empty lockages while still decreasing the overall travel times is highly welcomed. Although the runtimes of VNS1000 are higher the computation times start at 20 s for smaller instances and range up 5000 s for the largest instances.

The relative improvement rates for shifts of single ships, swaps, shift of two vessels, shifts of three vessels and removing empty lockages are 82 %, 37 %, 5 %, 0.02 % and 36 %, respectively.

8   Conclusions and Future Research

Within this work, we introduced the *Interdependent Lock Scheduling Problem* (ILSP) and provided a *Variable Neighborhood Search* (VNS) based approach for solving the ILSP heuristically. While different approaches exist for simpler versions of lock scheduling problems, the ILSP aims at finding an optimal lock schedule for multiple watergates along a river like the Danube. The ILSP is—to our best knowledge – not yet addressed in the academic literature. Based on real-world data provided by the Austrian waterway administration, we were able to show that an optimization approach helps in finding lock schedules resulting in reduced overall ship travel times. Furthermore, the number of (especially empty) lockages could be reduced without increasing the runtimes considerably. Since the ILSP is a highly complex problem, further research will include additional aspects such as considering fairness according to ship waiting times or considering a 2D-bin packing for ship placement in lock chambers.

Acknowledgements. We want to thank our project partners viadonau– Österreichische Wasserstraßen-Gesellschaft mbH and Zentralanstalt für Meteorologie und Geodynamik for providing us valuable insights into meteorological as well as nautical processes and challenges related to them as well as providing us relevant data.

References

1. Asamer, J., Prandtstetter, M.: Estimating ship travel times on inland waterways. In: TRB (ed.) TRB Annual Meeting Compendium of Papers. 14–3020 (2014)
2. Coene, S., Spieksma, F.C.R.: The Lockmaster's problem. In: Caprara, A., Kontogiannis, S. (eds.) 11th Workshop on Algorithmic Approaches for Transportation Modelling, Optimization, and Systems. OpenAccess Series in Informatics (OASIcs), vol. 20, pp. 27–37. Schloss Dagstuhl–Leibniz-Zentrum fuer Informatik, Dagstuhl (2011)
3. Dolinsek, M., Hartl, S., Hartl, T., Hintergräber, B., Hofbauer, V., Hrusovsky, M., Maierbrugger, G., Matzner, B., Putz, L.M., Sattler, M., Schweighofer, J., Seemann, L., Simoner, M., Slavicek, D.: Handbuch der Donauschifffahrt. bmvit (2013)
4. European Commission: Action plan accompanying the european union strategy for the danube region (2010). http://www.danube-region.eu/component/edocman/action-plan-eusdr-pdf
5. European Commission: European union strategy for danube region (2010). http://www.danube-region.eu/component/edocman/communication-of-the-commission-eusdr-pdf
6. European Commission: Roadmap to a single european traffic transport area - towards a competitive and resource efficient transport system (2011). http://ec.europa.eu/transport/strategies/doc/2011_white_paper/white_paper_com(2011)_144_en.pdf
7. European Commission: EU transport in figures: statistical pocketbook 2013. Publications Office of the European Union (2013)
8. Hansen, P., Mladenović, N.: Variable neighborhood search: principles and applications. Eur. J. Oper. Res. **130**(3), 449–467 (2001)
9. Martinelli, D., Schonfeld, P.: Approximating delays at interdependent locks. J. Waterw. Port Coast. Ocean Eng. **121**(6), 300–307 (1995)
10. Ting, C.J., Schonfeld, P.: Efficiency versus fairness in priority control: Waterway lock case. J. Waterw. Port Coast. Ocean Eng. **127**(2), 82–88 (2001)
11. Ting, C.J., Schonfeld, P.: Effects of speed control on tow travel costs. J. Waterw. Port Coast. Ocean Eng. **125**(4), 203–206 (1999)
12. Verstichel, J.: The lock scheduling problem. Ph.D. thesis, KU Leuven Faculty of Engineering Science (2013)

13. Verstichel, J., Vanden Berghe, G.: A late acceptance algorithm for the lock scheduling problem. In: Voß, S., Pahl, J., Schwarze, S. (eds.) Logistik Management, pp. 457–478. Physica-Verlag HD, Heidelberg (2009)
14. viadonau: Locked-through vessel units (2014). http://www.donauschifffahrt.info/en/facts_figures/statistics/locked_through_vessel_units/

A Variable Neighborhood Search
for the Generalized Vehicle Routing Problem
with Stochastic Demands

Benjamin Biesinger[✉], Bin Hu, and Günther R. Raidl

Institute of Computer Graphics and Algorithms, Vienna University of Technology,
Favoritenstraße 9–11/1861, 1040 Vienna, Austria
{biesinger,hu,raidl}@ads.tuwien.ac.at

Abstract. In this work we consider the generalized vehicle routing problem with stochastic demands (GVRPSD) being a combination of the generalized vehicle routing problem, in which the nodes are partitioned into clusters, and the vehicle routing problem with stochastic demands, where the exact demands of the nodes are not known beforehand. It is an NP-hard problem for which we propose a variable neighborhood search (VNS) approach to minimize the expected tour length through all clusters. We use a permutation encoding for the cluster sequence and consider the preventive restocking strategy where the vehicle restocks before it potentially runs out of goods. The exact solution evaluation is based on dynamic programming and is very time-consuming. Therefore we propose a multi-level evaluation scheme to significantly reduce the time needed for solution evaluations. Two different algorithms for finding an initial solution and three well-known neighborhood structures for permutations are used within the VNS. Results show that the multi-level evaluation scheme is able to drastically reduce the overall run-time of the algorithm and that it is essential to be able to tackle larger instances. A comparison to an exact approach shows that the VNS is able to find an optimal or near-optimal solution in much shorter time.

Keywords: Generalized vehicle routing problem · Stochastic vehicle routing problem · Variable neighborhood search · Stochastic optimization

1 Introduction

The generalized vehicle routing problem with stochastic demands (GVRPSD) combines the vehicle routing problem with stochastic demands (VRPSD) with the generalized vehicle routing problem (GVRP) and is a stochastic combinatorial optimization problem.

In the GVRPSD we are given a weighted complete undirected graph $G = (V, E)$ with a set of nodes V and a set of edges E. The edges $(i, l) \in E$ are

This work is supported by the Austrian Science Fund (FWF) grant P24660-N23.

G. Ochoa and F. Chicano (Eds.): EvoCOP 2015, LNCS 9026, pp. 48–60, 2015.
DOI: 10.1007/978-3-319-16468-7_5

weighted with distances $d_{il} \geq 0$. The set of nodes is partitioned into m disjoint subsets or clusters $C = \{C_0, C_1, \ldots, C_m\}, C_0, \ldots, C_m \subseteq V$, such that $C_0 \cup C_1 \cup \cdots \cup C_m = V$. Node $v_0 \in V$ is a dedicated depot and the only node of cluster C_0. Each other cluster $C_j, \forall j = 1, \ldots, m$ has an associated demand ξ_j which is a random variable following a known discrete probability distribution, i.e., we know for each cluster C_j the probability $p_{jk} = P(\xi_j = k)$ that cluster j has a demand of $k \geq 0$. Furthermore, we are given a vehicle with a limited capacity Q. For avoiding the necessity of multiple visits we assume that $p_{jk} = 0, \forall j = 1, \ldots, m, \forall k > Q$. The aim is to find a route visiting exactly one node from each cluster $C_1, \ldots, C_m$ exactly once and thereby distributing goods according to the clusters' actual demands. The current load of the vehicle decreases each time a cluster demand is satisfied but is refilled to Q each time it returns to the depot. The amount of how much the load gets decreased by visiting cluster j is dependent on the actual realization of ξ_j, which becomes known only upon arrival. Possibly, the vehicle will get empty, i.e., the current load becomes zero and the vehicle has to restock at the depot before continuing the route. Such an event is called a *stockout*. A common approach for solving such stochastic optimization problems is the use of a-priori routes, which has already been used for several probabilistic problems [3, 4, 13]. A-priori tours are planned before the actual realizations of the random variables are known but taking their probability distributions into account. The aim of the problem is to minimize the expected length of the tour under all possible a-posteriori tours.

In the literature there are several restocking strategies described for the VRPSD which can be adapted to the GVRPSD as well. The by far most common is the *standard restocking* policy [7, 12, 15, 20] where on each stockout the vehicle returns to the depot, refills its load, and continues its tour at the last visited node. Another method for handling stockouts is *re-optimization* [22]: whenever a stockout occurs the vehicle returns to the depot and then the tour through the remaining clusters is re-planned. Apart from the re-optimization approach, which can be problematic when implemented in practical applications, the *preventive restocking* policy [4, 16, 17, 24] is the most cost efficient.

In the preventive restocking method return trips to the depot can also be performed before an actual stockout occurs. It originates from the observation that a repeated visit of the same node after a restocking is usually expensive and can often be avoided if the restocking is done after servicing the preceding cluster. Especially on instances where the triangle inequalities hold, a restocking from the preceding cluster is always more cost efficient. Another advantage from using the preventive restocking strategy is that it is sufficient to plan one giant tour through all clusters when the problem is not further constrained. Yang et al. [24] proved this property for the VRPSD and it can be directly applied to the GVRPSD as well. Along with that proof they also proposed an evaluation procedure for the giant tour representation, which we will discuss in Sect. 3.

Like the GVRP this problem can be applied to the field of healthcare logistics, in which medical supplies are delivered to districts and the distributing company does not know beforehand how much supply is needed. If it does not matter to which hospital in each district these supplies are delivered this problem can be modelled as a GVRPSD. Another application domain is urban waste management, where refuse collecting vehicles gather waste from districts returning it to a central landfill site and the total amount of waste is not known beforehand.

In this work we describe a variable neighborhood search (VNS) [10] approach for the GVRPSD with preventive restocking. We use the giant tour representation and an evaluation procedure similar to the one used for the VRPSD [24], which is based on dynamic programming (DP). In addition to this solution evaluation procedure we also propose a multi-level evaluation scheme which iteratively approximates the quality of a solution candidate until it can be discarded as being inferior or the exact objective is obtained. In the next section the related work for this problem is discussed. Section 3 is dedicated to the solution representation and the associated evaluation procedure including the multi-level evaluation scheme. The actual variable neighborhood search and its operators are described in Sect. 4. We present computational results for this approach in Sect. 5 and draw conclusion of our work along with thoughts on future work in Sect. 6.

2   Related Work

To the best of our knowledge the GVRPSD has not been considered in the literature so far. However, there are many related problems which have been broadly discussed like the VRPSD, which is a special case of the GVRPSD but each cluster is a singleton. Especially the work by Yang et al. [24] and Bianchi et al. [4], which both used the preventive restocking strategy, has been inspiring to the approach for the GVRPSD presented here. Yang et al. [24] showed that planning multiple tours cannot lead to a better solution in the VRPSD. They also presented an evaluation procedure based on dynamic programming for the giant tour representation. The same evaluation is also used by Bianchi et al. [4] who developed several metaheuristics and an approximate delta evaluation for the VRPSD. The most recent work for the VRPSD utilizing the preventive restocking policy is by Marinakis et al. [17] who also applied the same evaluation procedure and presented a clonal selection algorithm. They reported promising results on their benchmark set and compared their algorithm to two versions of a particle swarm optimization [16], to a differential evolution, and a genetic algorithm.

Another related problem is the GVRP, which is the special case of the GVRPSD with deterministic demands. There are usually capacity or distance constraints on the used vehicles and therefore several routes have to be devised. There are several exact and heuristic solution approaches in the literature [1, 2, 8, 18, 19]. Since we are considering a giant tour approach the generalized traveling salesman problem (GTSP) is also closely related. The GTSP was originally introduced independently in [11, 21, 23]. In Fischetti et al. [5, 6] integer linear programming models are formulated and analyzed. For our construction heuristic we use one of their formulations.

3   Solution Representation and Evaluation

In the introduction we mentioned that it is sufficient to plan one giant tour through all clusters and therefore we use a solution representation based on the sequence of clusters that are visited in the tour. From this permutation of clusters we compute the expected cost using a DP algorithm based on the DP for the VRPSD [24]. We adapt it to the GVRPSD by also considering that we only have to visit one node per cluster. The worst case run-time complexity remains $\mathcal{O}(|V|Q^2)$.

First we describe the notations used in the DP. The function used for the recursion $f_{ij}(q)$ is defined for all $q = 0, \ldots, Q$, $j = 0, \ldots, m$, $i = 0, \ldots, |C_j|$ and can be interpreted as the remaining cost of the tour after servicing the i-th node of cluster j with the residual vehicle capacity q. We also define an auxiliary function $b_j(l)$ which returns the l-th node of cluster j. Let us assume that the clusters of the tour we want to evaluate are relabeled such that the tour is $t = (C_0, C_1, \ldots, C_m, C_0)$. Then the DP recursion is given by:

$$f_{ij}(q) = \min\{f_{ij}^p(q), f_{ij}^r(q)\}$$

$$f_{ij}^p(q) = \min_{l=0,\ldots,|C_{j+1}|} \left\{ d_{b_j(i),b_{j+1}(l)} + \sum_{k=0}^{q} f_{l,j+1}(q-k)p_{j+1,k} \right.$$

$$\left. + \sum_{k=q+1}^{Q} 2d_{b_{j+1}(l),0} + f_{l,j+1}(q+Q-k)p_{j+1,k} \right\}$$

$$f_{ij}^r(q) = d_{b_j(i),0} + \min_{l=0,\ldots,|C_{j+1}|} \left\{ d_{0,b_{j+1}(l)} + \sum_{k=0}^{Q} f_{l,j+1}(Q-k)p_{j+1,k} \right\}$$

$$\forall q = 0, \ldots, Q, j = 0, \ldots, m, i = 0, \ldots, |C_j|$$

and the boundary condition

$$f_{im}(q) = d_{b_m(i),0}, \quad \forall q = 0, \ldots, Q, \; i = 0, \ldots, |C_m|$$

The basic principle of this recursion is that for each node i and each vehicle load q it is computed if it is more cost-efficient to proceed directly to the next cluster which costs $f_{ij}^p(q)$ or to make a preventive restock which costs $f_{ij}^r(q)$. The total expected cost of the tour t is then given by the value of $f_{0,0}(Q)$. For our VNS such an expensive solution evaluation is inconvenient for larger instances with a large vehicle capacity. In the next section we describe a method to potentially reduce the run-time of the solution evaluation within the VNS framework, which can also be applied to other metaheuristics.

3.1   Multi-level Evaluation Scheme

In this section we describe a multi-level evaluation scheme (ML-ES) to iteratively estimate the exact objective value of a solution candidate with increasing accuracy until we either know that it cannot be better than the best solution found so far or we know its exact value. The basic idea is to scale down both the vehicle capacity and the probability distribution of the demand for each cluster accordingly. Since the time needed for the solution evaluation is quadratically dependent on Q, a large performance gain in terms of run-time is expected when Q is decreased.

In our ML-ES there are $\log_2 Q$ levels of approximation, where level 0 is the exact evaluation and $\log_2 Q$ is the roughest approximation level. Starting with level 0, increasing the level by one means to scale down the vehicle capacity Q and all demand distributions p_{jk} by a factor of two. We introduce a new vehicle capacity Q^i and new probabilities p^i_{jk} subject to level i, which are defined in the following way:

$$Q^0 = Q \tag{1}$$

$$p^0_{jk} = p_{jk} \qquad\qquad \forall j = 1,\ldots,|C|,\ k = 0,\ldots,Q \tag{2}$$

$$Q^i = \left\lceil \frac{Q^{i-1}}{2} \right\rceil \qquad\qquad \forall i = 1,\ldots,\lceil\log_2 Q\rceil \tag{3}$$

$$p^i_{jk} = p^{i-1}_{j,2k} + p^{i-1}_{j,2k+1} \qquad\qquad \forall j = 1,\ldots,|C|,\ k = 0,\ldots,Q^i, \tag{4}$$
$$\forall i = 1,\ldots,\lceil\log_2 Q\rceil$$

Figure 1 shows exemplarily for one cluster how the probability distribution for the demand changes at each level.

Not only is level $i \geq 1$ an approximation, but its objective value is also a lower bound for the objective value of the preceding level $i - 1$, which we will show next.

Lemma 1. *With increasing level i the ratio of the scaled expected demand of each cluster to the vehicle capacity Q^i is non-increasing.*

Proof. We have to show for each cluster j and each demand $0 \leq k \leq Q$ that

$$\frac{\sum_{k=0}^{Q^i} k p^i_{jk}}{Q^i} \leq \frac{\sum_{k=0}^{Q^{i-1}} k p^{i-1}_{jk}}{Q^{i-1}}, \qquad \forall i = 1,\ldots,\lceil\log_2 Q\rceil$$

is valid. Suppose to the contrary that for one cluster j and one demand k the following holds:

$$\frac{k p^i_{jk}}{Q^i} = \frac{k(p^{i-1}_{j,2k} + p^{i-1}_{j,2k+1})}{\frac{Q^{i-1}}{2}} > \frac{2k p^{i-1}_{j,2k} + (2k+1)p^{i-1}_{j,2k+1}}{Q^{i-1}} = \frac{k p^{i-1}_{jk}}{Q^{i-1}}$$

$$2k p^{i-1}_{j,2k} + 2k p^{i-1}_{j,2k+1} > 2k p^{i-1}_{j,2k} + 2k p^{i-1}_{j,2k+1} + p^{i-1}_{j,2k+1}$$

$$0 > p^{i-1}_{j,2k+1}$$

Obviously, this is a contradiction because all probabilities must be non-negative. Therefore, as no $\frac{k p^i_{jk}}{Q^i}$ can be larger than $\frac{k p^{i-1}_{jk}}{Q^{i-1}}$ for any cluster j this also holds for the sum over all demands, which proves the Lemma. $\square$

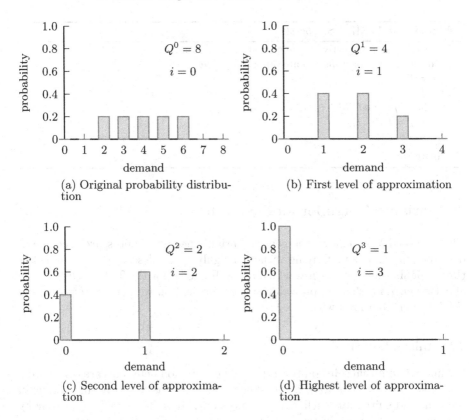

(a) Original probability distribution

(b) First level of approximation

(c) Second level of approximation

(d) Highest level of approximation

Fig. 1. An exemplary demand probability distribution and its different levels of approximation.

Theorem 1. *Let $c^i(t)$ be the objective value of a tour t on approximation level i. For each tour t it holds that $c^i(t) \leq c^{i-1}(t), \forall i = 1, \ldots, \lceil \log_2 Q \rceil$.*

Proof. Due to Lemma 1 it follows that the total expected relative demand of all clusters on level i is smaller or equal to that of level $i - 1$. So we can possibly service more customers before a restocking is needed and therefore the resulting objective $c^i(t)$ value is a lower bound to the exact objective value $c^0(t)$ and to the objective value at the preceding level $c^{i-1}(t)$. □

Algorithm 1 describes our ML-ES in pseudocode, where $DP(t, i)$ executes the DP described above with the scaled vehicle capacity and probability distributions according to (1–4). Algorithm 1 returns either the exact objective value of t if $DP(t, 0)$ is executed or a lower bound to the exact value otherwise. In the latter case the solution candidate can immediately be discarded because we know that it cannot be better than the best solution found so far.

Algorithm 1. ML-ES(t, *bestObj*)

Input : tour t, objective value of best solution found so far *bestObj*
Output: exact or approximate objective value
$obj = 0$;
$i = \lceil \log_2 Q \rceil$;
while $obj < bestObj \land i \geq 0$ **do**
$\quad\lfloor \quad obj = DP(t, i)$;
$\qquad i = i - 1$;
return obj;

---

4   Variable Neighborhood Search

The proposed VNS follows the general variable neighborhood search scheme as described in [9]. The underlying variable neighborhood descent (VND) considers three neighborhood structures which are well-known for the TSP: *1-shift*, *2-opt*, and *Or-opt*. As a shaking procedure for diversification we perform k^2 moves in the k-th neighborhood with $k = 1, \ldots, 3$.

4.1   Initial Solution

For finding an initial solution two types of construction heuristics are considered. The first, *farthest insertion*, is well-known for the classical traveling salesman problem and suited for Euclidean instances only. It builds iteratively a tour by starting at the depot cluster and inserting the cluster which is farthest away from the last inserted cluster at the best possible position. For that purpose we have to define distances between clusters, which is done by computing the geometric centers of clusters by taking the average of the x- and y-coordinates of its nodes. Then the distance between two clusters is the Euclidean distance between their centers.

An alternative but much more time-consuming method for finding a starting solution is solving the GTSP relaxation of the problem. From the solution of the GTSP relaxation we extract the cluster sequence which is then our initial solution. The GTSP is solved exactly by using a branch-and-cut algorithm with CPLEX and the E-GTSP formulation described in [6].

4.2   Neighborhood Structures

Three types of neighborhood structures are used in the VND part, which are searched with a best improvement step function in the order they are described here.

1-shift: A cluster is shifted to another position of the tour.

2-opt: A subsequence of the tour is inverted.

Or-opt: First two, then three consecutive clusters are shifted to another position of the tour. Note that Or-opt usually starts by shifting only one cluster in the tour but we covered this case by our first neighborhood structure and omit it here.

5   Computational Results

The VNS is implemented in C++ and for the GTSP starting solution CPLEX in version 12.5 is used. All our tests were carried out on a single core of an Intel Xeon with 2.53 GHz and 3GB RAM. We created Euclidean instances for the GVRPSD[1] based on instances for the GVRP [2] by assigning the original demand values to be the expected demands and deciding independently at random for each cluster if it is a *low spread* or a *high spread* cluster. For low spread clusters the set of possible demands is $\pm 10\%$ of the expected value and for the high spread it is $\pm 30\%$. All of these demand values are considered equally likely, so we assumed a uniform distribution over these values. We do not consider demand values smaller than zero or larger than Q. Due to space limitations the numerical results presented in this section are based on a representative selection of 37 instances out of the originally proposed 158 instances. These instances are selected such that a comparison to an existing exact method is possible. The full result tables can be found at our website (see Footnote 1).

In the following tables the column *Instance* contains the name of the instance, followed by the number of nodes n and the number of clusters m. Additionally the expected number of restocks $E[nr]$ are presented. Then the results for the different configurations are given with their (average) objective values, their standard deviations (sd) if applicable and either the total run-time in seconds $t[s]$ for the deterministic configurations or the average time when the best solution is identified $t^*[s]$.

First we compare the results of the different starting solutions, farthest insertion (FI) and GTSP, with a subsequent VND using the neighborhood structures and the order described in Sect. 4.2. Table 1 shows the (deterministic) numerical results for these configurations. Additionally it contains a third configuration where the ML-ES is used along with the GTSP starting solution.

The results indicate that both starting solutions produce similarly good results for the instances with up to 75 nodes. However, when considering larger instances with 76 nodes and more, FI is not competitive anymore. When starting from an inferior solution produced by FI the VND needs too much time and could not even be completed within the time limit of 10000 s. When comparing run-time we also see the advantage of using the GTSP over the FI; for most of the instances, especially for the larger ones, it pays off to invest more time to get a better starting solution so that the subsequent VND does not need so many iterations. We also applied the ML-ES to the GTSP + VND configuration and we observe a huge drop in run-time. It is clear that the resulting solution is the same as in the GTSP + VND configuration but the run-time could be reduced

---

[1] https://www.ads.tuwien.ac.at/w/Research/Problem_Instances#Generalized_ Vehicle_Routing_Problem_with_Stochastic_Demands.

Table 1. Results for the different configurations of the VND.

Instance	n	m	$E[nr]$	FI + VND		GTSP + VND		GTSP + VND + ML-ES	
				obj	*t[s]*	*obj*	*t[s]*	*obj*	*t[s]*
P-n19-k2-C7-V1	19	7	0,71	**112,105**	5	**112,105**	3	**112,105**	<1
P-n20-k2-C7-V1	20	7	0,68	**117,306**	4	**117,306**	<1	**117,306**	<1
P-n21-k2-C7-V1	21	7	0,64	**117,071**	3	**117,071**	<1	**117,071**	<1
P-n22-k2-C8-V1	22	8	0,73	**111,194**	10	**111,194**	5	**111,194**	<1
B-n31-k5-C11-V2	31	11	1,38	**355,729**	25	**355,729**	16	**355,729**	5
A-n32-k5-C11-V2	32	11	1,39	**386,909**	20	388,597	10	388,597	2
A-n33-k5-C11-V2	33	11	1,52	**318,028**	17	**318,028**	15	**318,028**	3
A-n33-k6-C11-V2	33	11	1,91	**367,629**	23	**367,629**	16	**367,629**	4
A-n34-k5-C12-V2	34	12	1,66	**419,124**	29	**419,124**	25	**419,124**	4
B-n34-k5-C12-V2	34	12	1,34	**363,089**	33	**363,089**	13	**363,089**	5
B-n35-k5-C12-V2	35	12	1,54	**501,470**	32	**501,470**	14	**501,470**	6
A-n36-k5-C12-V2	36	12	1,34	404,579	30	**399,905**	23	**399,905**	7
A-n37-k5-C13-V2	37	13	1,43	**359,133**	45	**359,133**	20	**359,133**	3
A-n37-k6-C13-V2	37	13	1,95	467,266	31	**430,987**	32	**430,987**	7
A-n38-k5-C13-V2	38	13	1,71	**371,795**	57	**371,795**	20	**371,795**	2
B-n38-k6-C13-V2	38	13	1,93	**386,195**	55	389,241	27	389,241	7
A-n39-k5-C13-V2	39	13	1,48	390,400	47	**371,410**	20	**371,410**	8
A-n39-k6-C13-V2	39	13	1,83	**417,844**	43	**417,844**	40	**417,844**	8
B-n39-k5-C13-V2	39	13	1,45	**281,482**	50	**281,482**	<1	**281,482**	<1
P-n40-k5-C14-V2	40	14	1,51	**214,775**	175	**214,753**	49	**214,753**	4
B-n41-k6-C14-V2	41	14	1,82	**404,261**	93	**404,261**	34	**404,261**	11
B-n43-k6-C15-V2	43	15	1,81	394,529	74	**347,650**	33	**347,650**	6
A-n44-k6-C15-V2	44	15	2,00	**505,129**	105	508,981	51	508,981	15
B-n45-k5-C15-V2	45	15	1,51	**419,613**	116	**419,613**	59	**419,613**	8
B-n45-k6-C15-V2	45	15	1,96	**358,989**	72	**358,989**	83	**358,989**	31
P-n45-k5-C15-V2	45	15	1,61	**239,568**	172	**239,357**	94	**239,357**	6
A-n45-k6-C15-V3	45	15	2,09	**478,219**	105	**478,219**	56	**478,219**	12
A-n45-k7-C15-V3	45	15	2,06	516,508	94	**488,017**	99	**488,017**	34
A-n46-k7-C16-V3	46	16	2,08	**465,624**	209	471,980	82	471,980	16
A-n48-k7-C16-V3	48	16	2,13	474,210	150	**462,548**	95	**462,548**	35
A-n53-k7-C18-V3	53	18	2,09	450,973	268	**443,875**	97	**443,875**	16
A-n54-k7-C18-V3	54	18	2,19	507,805	201	**490,544**	134	**490,544**	41

(*Continued*)

Table 1. *Continued*

Instance	n	m	$E[nr]$	FI + VND		GTSP + VND		GTSP + VND + ML-ES	
				obj	*t[s]*	*obj*	*t[s]*	*obj*	*t[s]*
A-n55-k9-C19-V3	55	19	2,75	475,919	292	**474,048**	114	**474,048**	15
A-n60-k9-C20-V3	60	20	2,80	**614,515**	517	620,897	361	620,897	117
P-n76-k4-C26-V2	76	26	1,33	461,753	>10000	**310,397**	4312	**310,397**	58
P-n76-k5-C26-V2	76	26	1,67	373,937	>10000	**310,397**	3748	**310,397**	56
P-n101-k4-C34-V2	101	34	1,25	992,679	>10000	**371,926**	9979	**371,926**	397

Table 2. Comparison of the proposed VNS with an exact integer L-shaped method.

Instance	n	m	$E[nr]$	L-shaped			VNS + GTSP + ML-ES		
				obj	*gap*	*t[s]*	$\overline{obj}$	*sd*	$t^*[s]$
P-n19-k2-C7-V1	19	7	0,71	**112,105**	0,0 %	<1	**112,105**	0,00	<1
P-n20-k2-C7-V1	20	7	0,68	**117,306**	0,0 %	<1	**117,306**	0,00	<1
P-n21-k2-C7-V1	21	7	0,64	**117,071**	0,0 %	<1	**117,071**	0,00	<1
P-n22-k2-C8-V1	22	8	0,73	**111,194**	0,0 %	<1	**111,194**	0,00	<1
B-n31-k5-C11-V2	31	11	1,38	**355,729**	11,1 %	>7200	**355,729**	0,00	5
A-n32-k5-C11-V2	32	11	1,39	**386,909**	0,0 %	1331	**386,909**	0,00	25
A-n33-k5-C11-V2	33	11	1,52	**318,028**	0,0 %	374	**318,028**	0,00	2
A-n33-k6-C11-V2	33	11	1,91	**364,589**	0,0 %	598	**364,589**	0,00	27
A-n34-k5-C12-V2	34	12	1,66	**419,124**	0,0 %	3719	**419,124**	0,00	4
B-n34-k5-C12-V2	34	12	1,34	**363,089**	18,1 %	>7200	**363,089**	0,00	4
B-n35-k5-C12-V2	35	12	1,54	501,470	26,3 %	>7200	**501,450**	0,11	6
A-n36-k5-C12-V2	36	12	1,34	**399,905**	9,4 %	>7200	**399,905**	0,00	7
A-n37-k5-C13-V2	37	13	1,43	**359,133**	0,0 %	98	**359,133**	0,00	3
A-n37-k6-C13-V2	37	13	1,95	434,865	18,5 %	>7200	**430,987**	0,00	7
A-n38-k5-C13-V2	38	13	1,71	**371,795**	0,0 %	588	**371,795**	0,00	2
B-n38-k6-C13-V2	38	13	1,93	**386,195**	12,2 %	>7200	388,734	1,15	8
A-n39-k5-C13-V2	39	13	1,48	**371,410**	7,6 %	>7200	**371,410**	0,00	8
A-n39-k6-C13-V2	39	13	1,83	**417,844**	5,6 %	>7200	**417,844**	0,00	8
B-n39-k5-C13-V2	39	13	1,45	**281,482**	0,0 %	282	**281,482**	0,00	<1
P-n40-k5-C14-V2	40	14	1,51	**214,753**	0,0 %	392	**214,753**	0,00	4
B-n41-k6-C14-V2	41	14	1,82	408,977	16,7 %	>7200	**404,261**	0,00	10
B-n43-k6-C15-V2	43	15	1,81	347,650	19,7 %	>7200	**347,650**	0,00	6
A-n44-k6-C15-V2	44	15	2,00	509,254	16,2 %	>7200	**508,981**	0,00	15
B-n45-k5-C15-V2	45	15	1,51	**419,613**	3,6 %	>7200	419,096	1,05	8
B-n45-k6-C15-V2	45	15	1,96	367,730	23,6 %	>7200	**358,989**	0,00	31

(Continued)

Table 2. (*Continued*)

Instance	n	m	E[nr]	L-shaped			VNS + GTSP + ML-ES		
				obj	gap	t[s]	$\overline{obj}$	sd	t*[s]
P-n45-k5-C15-V2	45	15	1,61	**239,357**	4,6 %	>7200	**239,357**	0,00	6
A-n45-k6-C15-V3	45	15	2,09	478,265	16,2 %	>7200	**478,219**	0,00	12
A-n45-k7-C15-V3	45	15	2,06	491,539	30,6 %	>7200	**488,017**	0,00	34
A-n46-k7-C16-V3	46	16	2,08	**465,624**	19,0 %	>7200	471,539	0,51	16
A-n48-k7-C16-V3	48	16	2,13	469,690	28,7 %	>7200	**462,548**	0,00	35
A-n53-k7-C18-V3	53	18	2,09	**443,873**	13,6 %	>7200	443,875	0,00	16
A-n54-k7-C18-V3	54	18	2,19	500,349	28,6 %	>7200	**490,544**	0,00	41
A-n55-k9-C19-V3	55	19	2,75	483,997	21,7 %	>7200	**474,048**	0,00	15
A-n60-k9-C20-V3	60	20	2,80	623,528	35,6 %	>7200	**617,575**	4,98	118
P-n76-k4-C26-V2	76	26	1,33	**310,397**	6,1 %	>7200	**310,397**	0,00	55
P-n76-k5-C26-V2	76	26	1,67	**310,397**	5,8 %	>7200	**310,397**	0,00	53
P-n101-k4-C34-V2	101	34	1,25	**371,926**	5,7 %	>7200	**371,926**	0,00	379

substantially. Only by using the ML-ES the VND is about 10 times faster on average with a peak speedup factor of 75 for instance P-n76-k4-C26-V2. During our tests when a solution is evaluated using ML-ES the procedure could be terminated in the top 30 % of the approximation levels where the acceleration is the largest.

Next we show how average results over 30 independent runs of the VNS with the GTSP starting solution and the ML-ES compares to an exact algorithm. The exact algorithm is the integer L-shaped method [14] applied to the GVRPSD which is a two-level approach based on a mixed integer programming model for the GTSP. Within a branch-and-cut framework it iteratively adds cuts generated in the lower level setting a lower bound on the restocking costs in the upper level. Due to space limitation this method is not described here in more detail. For the exact algorithm the optimality gap (*gap*) and the time needed is stated in the table.

Table 2 shows the numerical comparison with the exact method. The high optimality gaps on the medium to large instances show that the GVRPSD is a hard problem but on the instances where the L-shaped method is able to find a proven optimal solution the VNS also finds it in substantially less time. In the extreme case of instance A-n34-k5-C12-V2 the VNS found an optimal solution in all 30 runs about 865 times faster than the exact algorithm. However, since the L-shaped method guarantees the optimality of the solution we cannot directly compare the run-time of these algorithms. Our tests further showed that in the VNS the ML-ES can be terminated within the top 4 % of the approximation levels which is even better than for the VND.

6   Conclusions and Future Work

In this work a variable neighborhood search approach for the generalized vehicle routing problem with stochastic demands under the preventive restocking policy is presented. The problem has not yet been considered so far in the literature although many real world problems can be modelled in this way. An initial attempt to solve this hard stochastic combinatorial optimization problem was made. Therefore, concepts from both, the related VRPSD and the GTSP, are used. The solution representation and the solution evaluation method that had proved to work well for the VRPSD were adapted to the GVRPSD. On top of that a multi-level evaluation scheme was used to substantially reduce the time needed for evaluating a solution candidate. The computational results show that the VNS with the GTSP initial solution and the ML-ES is able to find optimal or near-optimal solutions in short times. Future work can include the development and improvement of the exact algorithm for the GVRPSD to get optimal solutions to more instances. In a following step also the combination of the techniques presented here, especially the ML-ES, and the methods used for solving this problem exactly should be investigated. Although the ML-ES is primarily developed for the GVRPSD applications to other similar problems like the VRPSD are also possible and could lead to a significant performance gain.

References

1. Afsar, H.M., Prins, C., Santos, A.C.: Exact and heuristic algorithms for solving the generalized vehicle routing problem with flexible fleet size. Int. Trans. Oper. Res. **21**(1), 153–175 (2014)
2. Bektaş, T., Erdoğan, G., Røpke, S.: Formulations and branch-and-cut algorithms for the generalized vehicle routing problem. Trans. Sci. **45**(3), 299–316 (2011)
3. Bertsimas, D.J.: Probabilistic combinatorial optimization problems. Ph.D. thesis, Massachusetts Institute of Technology (1988)
4. Bianchi, L., Birattari, M., Chiarandini, M., Manfrin, M., Mastrolilli, M., Paquete, L., Rossi-Doria, O., Schiavinotto, T.: Hybrid metaheuristics for the vehicle routing problem with stochastic demands. J. Math. Model. Algorithms **5**(1), 91–110 (2006)
5. Fischetti, M., Salazar González, J.J., Toth, P.: The symmetric generalized traveling salesman polytope. Networks **26**(2), 113–123 (1995)
6. Fischetti, M., Salazar González, J.J., Toth, P.: A branch-and-cut algorithm for the symmetric generalized traveling salesman problem. Oper. Res. **45**(3), 378–394 (1997)
7. Gendreau, M., Laporte, G., Séguin, R.: A tabu search heuristic for the vehicle routing problem with stochastic demands and customers. Oper. Res. **44**(3), 469–477 (1996)
8. Hà, M.H., Bostel, N., Langevin, A., Rousseau, L.M.: An exact algorithm and a metaheuristic for the generalized vehicle routing problem with flexible fleet size. Comput. Oper. Res. **43**, 9–19 (2014)
9. Hansen, P., Mladenović, N.: Variable neighborhood search. In: Glover, F., Kochenberger, G. (eds.) Handbook of Metaheuristics. International Series in Operations Research & Management Science, vol. 57, pp. 145–184. Springer, US (2003)

10. Hansen, P., Mladenović, N., Moreno Pérez, J.: Variable neighbourhood search: methods and applications. Ann. Oper. Res. **175**(1), 367–407 (2010)
11. Henry-Labordere: The record balancing problem: a dynamic programming solution of the generalized traveling salesman problem. RAIRO Oper. Res. **B2**, 43–49 (1969)
12. Hjorring, C., Holt, J.: New optimality cuts for a single vehicle stochastic routing problem. Ann. Oper. Res. **86**, 569–584 (1999)
13. Jaillet, P.: Probabilistic traveling salesman problems. Ph.D. thesis, Massachusetts Institute of Technology (1985)
14. Laporte, G., Louveaux, F.V., van Hamme, L.: An integer L-shaped algorithm for the capacitated vehicle routing problem with stochastic demands. Oper. Res. **50**(3), 415–423 (2002)
15. Laporte, G., Louveaux, F.: Solving stochastic routing problems with the integer L-shaped method. In: Crainic, T., Laporte, G. (eds.) Fleet Management and Logistics, pp. 159–167. Springer, New York (1998)
16. Marinakis, Y., Iordanidou, G.R., Marinaki, M.: Particle swarm optimization for the vehicle routing problem with stochastic demands. Appl. Soft Comput. **13**(4), 1693–1704 (2013)
17. Marinakis, Y., Marinaki, M., Migdalas, A.: A hybrid clonal selection algorithm for the vehicle routing problem with stochastic demands. In: Pardalos, P.M., Resende, M.G., Vogiatzis, C., Walteros, J.L. (eds.) LION 2014. LNCS, vol. 8426, pp. 258–273. Springer, Heidelberg (2014)
18. Pop, P.C., Fuksz, L., Marc, A.H.: A variable neighborhood search approach for solving the generalized vehicle routing problem. In: Polycarpou, M., de Carvalho, A.C.P.L.F., Pan, J.-S., Woźniak, M., Quintian, H., Corchado, E. (eds.) HAIS 2014. LNCS, vol. 8480, pp. 13–24. Springer, Heidelberg (2014)
19. Pop, P.C., Kara, I., Marc, A.H.: New mathematical models of the generalized vehicle routing problem and extensions. Appl. Math. Model. **36**(1), 97–107 (2012)
20. Rei, W., Gendreau, M., Soriano, P.: A hybrid monte carlo local branching algorithm for the single vehicle routing problem with stochastic demands. Transp. Sci. **44**(1), 136–146 (2010)
21. Saskena, J.: Mathematical model of scheduling clients through welfare agencies. J. Can. Oper. Res. Soc. **8**, 185–200 (1970)
22. Secomandi, N., Margot, F.: Reoptimization approaches for the vehicle-routing problem with stochastic demands. Oper. Res. **57**(1), 214–230 (2009)
23. Srivastava, S.S., Kumar, S., Carg, R.C., Sen, P.: Generalized traveling salesman problem through n sets of nodes. Can. Oper. Res. Soc. J. **7**, 97–101 (1969)
24. Yang, W.H., Mathur, K., Ballou, R.H.: Stochastic vehicle routing problem with restocking. Transp. Sci. **34**(1), 99–112 (2000)

An Iterated Local Search Algorithm for Solving the Orienteering Problem with Time Windows

Aldy Gunawan$^{(\boxtimes)}$, Hoong Chuin Lau, and Kun Lu

School of Information Systems, Singapore Management University,
80 Stamford Road, Singapore 178902, Singapore
{aldygunawan,hclau,kunlu}@smu.edu.sg

Abstract. The Orienteering Problem with Time Windows (OPTW) is a variant of the Orienteering Problem (OP). Given a set of nodes including their scores, service times and time windows, the goal is to maximize the total of scores collected by a particular route considering a predefined time window during which the service has to start. We propose an Iterated Local Search (ILS) algorithm to solve the OPTW, which is based on several LOCALSEARCH operations, such as SWAP, 2-OPT, INSERT and REPLACE. We also implement the combination between ACCEPTANCECRITERION and PERTURBATION mechanisms to control the balance between diversification and intensification of the search. In PERTURBATION, SHAKE strategy is introduced. The computational results obtained by our proposed algorithm are compared against optimal solutions or best known solution values obtained by state-of-the-art algorithms. We show experimentally that our proposed algorithm is effective on well-known benchmark instances available in the literature. It is also able to improve the best known solution of some benchmark instances.

Keywords: Orienteering problem · Time windows · Iterated local search

1 Introduction

The Orienteering Problem (OP) was first introduced by Tsiligirides in [1]. The main objective is to select a subset of nodes and define the sequence of selected nodes so that the total collected score is maximized while the maximum total travel time (time budget given) is not exceeded. The recent survey of real-life applications of the OP and its variants is presented by Vansteenwegen et al. in [2].

The Orienteering Problem with Time Windows (OPTW) is a variant of the OP with time window constraints that arise in situations where nodes/locations have to be visited within a predefined time window specified by an earliest and a latest time into which the service has to start [3]. An early arrival to a particular node leads to waiting times, while a late arrival causes infeasibility. Given a set of nodes, each one with a score, the goal is to maximize the total of collected score by a particular route subject to a time budget and time window constraints. The OPTW can be extended to the Team Orienteering Problem with Time Windows (TOPTW) when the number of route considered is more than one route [4].

© Springer International Publishing Switzerland 2015
G. Ochoa and F. Chicano (Eds.): EvoCOP 2015, LNCS 9026, pp. 61–73, 2015.
DOI: 10.1007/978-3-319-16468-7_6

In this paper, an Iterated Local Search (ILS) algorithm is proposed to solve the OPTW. The algorithm starts with generating an initial solution, which is constructed by inserting nodes subsequently into a route. A set of feasible candidate nodes to be inserted is created and the selection of a node to be inserted is based on roulette-wheel selection [5]. The initial solution is further improved by ILS. We consider components of ILS: LOCALSEARCH, PERTURBATION, and ACCEPTANCECRITERION. The LOCALSEARCH procedure involves several operations, such as SWAP, 2-OPT, REPLACE and INSERT.

In Sect. 2, we present the problem description and literature review of the OPTW. Section 3 is devoted to the proposed algorithm. Section 4 provides the computation results together with the analysis of the results. Section 5 concludes the paper and summarizes directions for further research.

2   Problem Description and Literature Review

The OPTW is defined as follows. Let us consider a set of nodes $N = \{1, 2, \cdots, n\}$ where each node $i \in N$ is associated with a score u_i and a service time T_i. The starting and end nodes are assumed to be nodes 1 and n, respectively; therefore, u_1, T_1, u_n, T_n are set to 0. The non-negative travel time between nodes i and j is represented as t_{ij}.

Each node i associates with a time window $[e_i, l_i]$, where e_i and l_i are the earliest and latest times allowed for starting service at node i. We assume that $e_1 = e_n = 0$ and $l_1 = l_n = T^{max}$. For mathematical formulations for the OPTW, we refer to [4,6]. The objective of the OPTW is to maximize the total collected score when visiting a subset of the nodes with respect to following constraints, as listed below:

- The route starts and ends at nodes 1 and n, respectively.
- Each node $i \in N$ is visited at most once.
- The service start time at node i is within a time window $[e_i, l_i]$.
- The time budget is limited by T^{max}.

The initial investigation of the OPTW has been presented by Kantor and Rosenwein in [6]. Since OPTW falls into NP-hard, a heuristic based on the tree heuristic was proposed. The experiments showed that the tree heuristic outperforms the insertion heuristic. Righini and Salani [7] proposed an exact optimization algorithm for the OPTW. The algorithm is based on dynamic programming with decremental state space relaxation. The result shows that there is no domination between the proposed algorithm and the other dynamic programming proposed by Boland et al. in [8] for solving benchmark instances. A new heuristic technique for the initialization of the critical vertex set has also been proposed in order to reduce the number of iterations and the amount of computing time required.

The Tourist Trip Design Problems (TTDP) can be formulated as the OPTW and the TOPTW [4]. A simple, fast and effective Iterated Local Search (ILS) was proposed to solve both problems. The proposed algorithm only combines insertion and shaking operations to generate the solutions. New data set was designed to analyse the performance of the proposed algorithm and to be used as a benchmark for further research. Montemanni and Gambardella [9] proposed a heuristic approach based on Ant Colony System (ACS). It includes a local search procedure by exchanging two subchains of nodes of the giant tour. Experimental results on benchmark instances have proven the effectiveness of the algorithm. For other related works with further improvement of benchmark instances' results, we can refer to [10,11].

A Simulated Annealing-based heuristic was proposed by Lin and Yu in [12] for solving both OPTW and TOPTW. Two different versions, fast SA (FSA) and slow SA (SSA), were developed in order to tailor two different scenarios. The former is mainly for the applications that need quick responses while the latter is more concerned about the quality of the solutions. The SSA heuristic is able to find 33 new best solutions. A heuristic based on a Variable Neighborhood Search (VNS) was proposed in order to tackle the OPTW and the TOPTW [3]. The idea of granularity that includes time constraints and profits in addition to pure distances is introduced. The proposed algorithm has been able to improve 25 best known solution values.

Hu and Lim [13] proposed an iterative framework which is based on three components: a local search procedure, a Simulated Annealing procedure and ROUTE RECOMBINATION. The first two components are used to explore the solution space and discover a set of routes. The last component which focuses on combining the routes to identify high quality solutions is included. 35 new best solutions are found and more than 83 % of instances with optimal solutions can be found.

3 Proposed Algorithm

This section presents the description of our proposed algorithm. The algorithm is started by generating an initial feasible solution using a greedy construction heuristic. The initial solution is further improved by Iterated Local Search (ILS). Components of ILS: LOCALSEARCH, PERTURBATION and ACCEPTANCECRITERION, are taken into consideration. The differences between our ILS and ILS proposed by Vansteenwegen et al. [4] would be described below.

3.1 Greedy Construction Heuristic

The greedy construction heuristic builds an initial solution from scratch. The idea is to insert a node subsequently to a route until no more feasible insertion can be found. A node insertion is feasible if all scheduled nodes after the insertion still satisfy their respective time windows and the total spent time does not exceed T^{max}.

Let N' and N^* be the sets of unscheduled and scheduled nodes respectively $(N' \cup N^* = N)$. The greedy construction heuristic is outlined in Algorithm 1. N^* is initialized by nodes 1 and n, while N' consists of the remaining unscheduled nodes. S_0 represents the current feasible solution obtained so far, represented as a vector $(1 \times |N^*|)$.

Let F be the set of feasible candidate nodes to be inserted. F is generated iteratively in order to store feasible candidate unscheduled nodes to be inserted. The idea of generating F is summarized in Algorithm 2. P is denoted as the set of all positions of a route. We examine all possibilities of inserting an unscheduled node in position $p \in P$. Each element in F, which represents a feasible insertion of node n in position p of a route, is represented as $\langle n, p \rangle$. For each possible insertion, we calculate the benefit of insertion $ratio_{n,p}$ by using Eq. (1). $\Delta_{n,p}$ represents the difference between the total time spent before and after the insertion of node n in position p. For example, if the total time spent before the insertion of node n in position p is 700 time units and the total time spent after the insertion is increased to 720 time units, the value of $\Delta_{n,p}$ is $720 - 700 = 20$ time units. All elements would be sorted in descending order based on $ratio_{n,p}$ values and we only keep f elements in F and remove the rest.

Algorithm 1. CONSTRUCTION (N)

$N^* \leftarrow$ nodes 1 and n
$N' \leftarrow N\backslash$ nodes 1 and n
Initialize $S_0 \leftarrow N^*$
$F \leftarrow$ UPDATEF(N')
while $F \neq \emptyset$ **do**
   $\langle n^*, p^* \rangle \leftarrow$ SELECT(F)
   $S_0 \leftarrow \langle n^*, p^* \rangle$
   $N' \leftarrow N' \setminus \{n^*\}$
   $N^* \leftarrow N^* \cup \{n^*\}$
   $F \leftarrow$ UPDATEF(N')
end while
return S_0

Algorithm 2. UPDATEF (N')

$F \leftarrow \emptyset$
for all $n \in N'$ **do**
   for all $p \in P$ **do**
      if insert node n in position p is feasible **then**
         calculate $ratio_{n,p}$
         $F \leftarrow F \cup \langle n, p \rangle$
      end if
   end for
end for
Sort all elements of F in descending order based on $ratio_{n,p}$
Select the best f number of elements of F and remove the rest
return F

Algorithm 3. SELECT (F)

$SumRatio \leftarrow 0$
for all $\langle n, p \rangle \in F$ **do**
   $SumRatio \leftarrow SumRatio + ratio_{n,p}$
end for
for all $\langle n, p \rangle \in F$ **do**
   $prob_{n,p} \leftarrow ratio_{n,p}/SumRatio$
end for
$U \leftarrow rand(0, 1)$
$AccumProb \leftarrow 0$
for all $\langle n, p \rangle \in F$ **do**
   $AccumProb \leftarrow AccumProb + prob_{n,p}$
   if $U \leq AccumProb$ **then**
     $\langle n^*, p^* \rangle \leftarrow \langle n, p \rangle$
     break
   end if
end for
return $\langle n^*, p^* \rangle$

$$ratio_{n,p} = \left(\frac{u_n^2}{\Delta_{n,p}} \right) \quad \forall n \in N', p \in P \tag{1}$$

If $F \neq \emptyset$, Algorithm 3 is run in order to select which $\langle n^*, p^* \rangle$ to be inserted. Each $\langle n, p \rangle$ corresponds to probability value $prob_{n,p}$. The probability is calculated by Eq. (2):

$$prob_{n,p} = \left(\frac{ratio_{n,p}}{\sum_{\langle i,j \rangle \in F} ratio_{i,j}} \right) \quad \forall n \in N', p \in P \tag{2}$$

Instead of always selecting an inserted node with the highest value of $ratio_{n,p}$ [4], our approach is different. Selecting $\langle n^*, p^* \rangle$ from F is based on roulette-wheel selection [5]. This method assumes that the probability of selection a particular $\langle n, p \rangle$ is proportional to the benefit of its insertion, $ratio_{n,p}$. A random number $U \sim [0, 1]$ is generated. The accumulative of probability values, $AccumProb$, is initially set to 0. We select a particular $\langle n^*, p^* \rangle$ and update the value of $AccumProb$ iteratively. This loop will be terminated when $(U \leq AccumProb)$ and the corresponding $\langle n^*, p^* \rangle$ is selected. S_0, N' and N^* will then be updated. The greedy construction heuristic is terminated when there is no further feasible insertion $(F = \emptyset)$.

Due to the time windows, the score of a node insertion is more relevant compared against the time consumption of an insertion. By removing the square, the obtained results are worse [4]. Therefore, the square of score is then applied in Eq. (1). Another main reason is by using the square of score, we increase the probability of selecting a particular node with a higher ratio (Eq. (2)) since the main objective is to maximize the collected score.

3.2   Iterated Local Search

Given the initial solution S_0 generated by the greedy construction heuristic, we propose an Iterated Local Search (ILS) algorithm to further improve the quality of S_0. Three components of ILS: PERTURBATION, LOCALSEARCH and ACCEPTANCECRITERION, are taken into consideration. Let S^* be the best found solution so far. The outline of ILS is presented in Algorithm 4.

Algorithm 4. ILS (N)

$S_0 \leftarrow$ CONSTRUCTION(N)
$S_0 \leftarrow$ LOCALSEARCH(S_0, N^*, N')
$S^* \leftarrow S_0$
NOIMPR $\leftarrow 0$
while TIMELIMIT has not been reached **do**
    $S_0 \leftarrow$ PERTURBATION(S_0, N^*, N')
    $S_0 \leftarrow$ LOCALSEARCH(S_0, N^*, N')
    if S_0 better than S^* **then**
        $S^* \leftarrow S_0$
        NOIMPR $\leftarrow 0$
    else
        NOIMPR $\leftarrow$ NOIMPR $+ 1$
    end if
    if (NOIMPR$+1$) MOD THRESHOLD1 $= 0$ **then**
        $S_0 \leftarrow S^*$
    end if
end while
return S^*

PERTURBATION is applied to S_0 in order to escape from local optima. In this paper, we implement SHAKE operation. The SHAKE operation is adopted from [4] with some modifications. During SHAKE operation, one or more nodes will be removed, which depends on two integer values. The first one indicates how many consecutive nodes to be removed (denoted as *cons*), while the second one indicates the first position of the removed nodes (denoted as *post*). If the last scheduled node is reached and there are still some nodes to be removed, we go back to the start node and include nodes after the start node. Both *cons* and *post* are initially set to 1. After each SHAKE operation, *post* is increased by *cons*. *cons* remains the same for a fixed number of consecutive iterations, e.g. 2 iterations and it is then increased by 1 subsequently. In [4], *cons* will always be increased by 1 for each iteration. If *post* is greater than the size of the smallest route, *post* is subtracted with the size of the smallest route in order to determine the new position. If *cons* is greater than the size of the largest route, or S^* is updated, *cons* is reset to one. Again, this differs from [4] where *cons* is set to 1 if *cons* is equal to $n/3$. After removing *cons* nodes, we generate F based on Algorithm 2 and select a node to be inserted using Algorithm 3. N' and N^* are then updated accordingly. This is repeated until $F = \emptyset$.

ILS proposed by Vansteenwegen et al. [4] only considers INSERT and SHAKE operations for generating the solutions. In our LOCALSEARCH, we consider four different operations that would be explained as follows. SWAP is applied by exchanging two scheduled nodes within a route. All possible combinations of selecting two different scheduled nodes are examined. SWAP is executed if it increases the remaining travel time and there is no constraint violation. 2-OPT is started by selecting two positions of two scheduled nodes. The sequence of scheduled nodes is reversed as long as there is no constraint violation and there is an improvement of the remaining travel time. This would be terminated if no further improvement in terms of total of remaining travel time.

INSERT is applied in order to insert one unscheduled node to a route. It is started by generating F based on Algorithm 2 and selecting node $i \in N'$ to be inserted using Algorithm 3. This is repeated until $F = \emptyset$. The idea is the same with the one introduced in the greedy construction heuristic. The last operation, REPLACE, tries to replace one scheduled node $i \in N^*$ with one unscheduled node $j \in N'$ with the highest score u_j. We then check each position p and examine whether selected node j can replace the node in position p. The feasibility of the solution and the improvement of total score are considered in this operation. Once this operation is successful, we continue with the next unscheduled node j with the second highest score u_j. Otherwise, the operation would be terminated.

ACCEPTANCE CRITERION is described as follows. The new local optimum solution is always accepted as the initial solution for the next run of local search. However, if there is no improvement of S^* obtained for a certain number of iterations, ((NOIMPR+1) MOD THRESHOLD1 = 0), the search is continued by applying an intensification strategy. This strategy focuses the search once again starting from the best found solution, S^* in order to improve the probability of hitting the global optimum. Finally, the entire algorithm will be run within the computational budget, TIMELIMIT.

4   Computational Experiments

4.1   Benchmarks and Experimental Setup

The test problems for the OPTW in the literature were initially proposed by Righini and Salani in [7], which are generated from Solomon's [14] and Cordeau et al.'s instances [15]. 48 Solomon's instances contain 100 nodes of series (c100, r100 and rc100). Cordeau et al.'s instances consists of 10 instances with different number of nodes, varying from 48 to 288 nodes (pr01–pr10). Those instances were designed for the Vehicle Routing Problem with Time Windows (VRPTW) and the Multi Depot Periodic VRPTW respectively. In this paper, we only concern with benchmark instances with the number of route = 1, which related to the OPTW problem. 37 additional instances were created [9]. 27 instances are converted from Solomon's dataset (c200, r200 and rc200) and 10 instances are converted from Cordeau et al.'s dataset (pr11–pr20).

Table 1. Estimation of single-thread performance [13].

Algorithm	Experimental environment	SuperPi	Estimate of single-thread performance
IterLS	Intel Core 2 with 2.5 gigahertz CPU, 3.45 gigabytes RAM	18.6	0.53
ACS*	Dual AMD Opteron 250 2.4 gigahertz CPU, 4 gigabytes RAM	Unknown	0.22
SSA	Intel Core 2 CPU, 2.5 gigahertz	18.6	0.53
GVNS	Intel Pentium (R) IV, 3 gigahertz CPU	44.3	0.22
I3CH	Intel Xeon E5430 CPU clocked at 2.66 gigahertz, 8 gigabytes RAM	14.7	0.67
ILS	Intel Core i7-4770 with 3.4 GHz processor, 16 gigabytes RAM	9.8	1

The experiments were carried out on a personal computer Intel Core i7 - 4770 with 3.4 GHz processor and 16 GB RAM. Vansteenwegen et al. [4] discussed the difficulty of solving the instances by a commercial solver (CPLEX). ILS was tested by performing 10 runs with different random seeds per each instance. The performances of the proposed ILS are compared to the state-of-the-art methods: Iterated Local Search (IterLS) [4], Ant Colony System (ACS) [9], Enhanced Ant Colony System (Enhanced ACS) [11], Slow Simulated Annealing (SSA) [12], Granular Variable Neighborhood Search (GVNS) [3] and Iterative Three-Component Heuristic (I3CH) [13]. Enhanced ACS [11] has empirically outperformed the original ACS [9]. In this paper, we refer to the results of both, whichever is better and denote them as ACS*.

For each instance, ACS* was executed in 5 runs whereas ILS and GVNS were executed 10 times. On the other hand, IterLS, SSA and I3CH were only executed once and reported only the best found solutions. For comparison purpose, the solutions of our ILS were compared against the best known solutions (BKs) of IterLS, ACS*, SSA, GVNS and I3CH. In order to ensure the fair comparisons, we refer to the same approach [13] to compare the speed of the computers used in obtaining the solutions, as shown in Table 1. SuperPi is a single-threaded program that computes the first 1 million digits of π of a particular processor. The comparability of processors used by ACS* and GVNS is shown in [10] since the SuperPi for ACS* is not available.

By setting the performance of our machine to be 1, we then estimated the single-thread performance of other processors by multiplying with the single-thread performance estimation (last column of Table 1). For the details, please refer to [13]. Among all algorithms, only ACS* used one hour of the computational time for each instance, while the rest use the number of iterations. In this paper, we are more concerned with solution quality, we then used ACS* as our reference. Instead of using 100 % of ACS*'s computational time, we only

Table 2. New best known solution values found by ILS.

Instance	Old BK	New BK	Instance	Old BK	New BK	Instance	Old BK	New BK
r203	1021	1026	r209	950	956	rc206	895	899
r204	1086	1093	r211	1046	1049	rc208	1053	1057
r208	1112	1113	rc202	936	938			

use 35 % of it. The computational time for each instance is then set to 35 % ×
0.22 × 3600 = 272 s using our processor. Based on the preliminary testing, the
following parameter values seem to have the best performance within a reasonable computational time: $f = 5$ and THRESHOLD1 = 10.

4.2 Experimental Results

Table 2 reports the new best known solutions (BKs) obtained by ILS. We dis-
covered 8 new best known solution values for Solomon's instances. Partial results
obtained by ILS on benchmark instances are reported in Table 3. We only report
the results of Solomon's instances due to space constraints. The detailed results
can be found online at http://centres.smu.edu.sg/larc/Orienteering-Problem-
Library.

Table 3 consists of two identical structure parts. The first column contains
the instance name, the second column reports the best known solution value BK
from references. The following three columns show maximum, average and minimum solution values obtained by our ILS. The "BG (%)" column provides the
best relative percentage deviation, which refers to the percentage gap between
BK and the best solution obtained by ILS. "AG (%)" provides the average
relative percentage deviation, which refers to the percentage gap between BK
and the average solution obtained by ILS. Finally, the last three columns show
maximum, average and minimum computational times required to obtain the
best found.

Take note that the optimal value is indicated in *italic* and the new BK
obtained by ILS is indicated in **bold**. There are still 27 and 12 instances of
Solomon's and Cordeau et al.'s instances where the optimal values are unknown.
ILS is able to obtain 41 out 56 (≈73.2 %) best known solutions (BKs) on
Solomon's instances. It also improved the best known solutions of 8 out 27
instances (≈30.0 %). For Cordeau et al.'s instances, 12 out 20 (≈60.0 %) BKs
can be found by ILS.

Table 4 summarizes the results of IterLS, ACS*, GVNS, SSA, I3CH and our
ILS results. The *numb* column provides the number of instances in a particular instance set. The table reports the average of AG for each instance set
($\overline{AG}(\%)$). However, IterLS, SSA and I3CH only reported their best known solution obtained; therefore, we report the average of BG ($\overline{BG}(\%)$) as well. The
best known solutions (BKs) were collected from IterILS, ACS*, SSA, GVNS
and I3CH results. The computational time ($\overline{CPU}$) for ACS* and ILS for one
particular instance set reports the average of time spent to obtain the best

Table 3. Detailed results of ILS on Solomon's instances

Instance	BK	ILS			BG(%)	AG(%)	CPU(seconds)		
		Max	Avg	Min			Max	Avg	Min
c101	320	320	320.0	320	0.0	0.0	0.5	0.2	0.0
c102	360	360	360.0	360	0.0	0.0	0.8	0.3	0.0
c103	400	400	400.0	400	0.0	0.0	0.4	0.2	0.1
c104	420	420	420.0	420	0.0	0.0	1.0	0.4	0.1
c105	340	340	340.0	340	0.0	0.0	1.3	0.4	0.1
c106	340	340	340.0	340	0.0	0.0	0.9	0.5	0.1
c107	370	370	370.0	370	0.0	0.0	0.4	0.1	0.0
c108	370	370	370.0	370	0.0	0.0	0.9	0.5	0.1
c109	380	380	380.0	380	0.0	0.0	25.4	6.8	0.6
r101	*198*	198	198.0	198	0.0	0.0	0.4	0.1	0.0
r102	286	286	286.0	286	0.0	0.0	0.5	0.2	0.0
r103	293	293	293.0	293	0.0	0.0	3.9	1.4	0.2
r104	303	303	303.0	303	0.0	0.0	6.2	1.5	0.1
r105	247	247	247.0	247	0.0	0.0	1.3	0.7	0.0
r106	293	293	293.0	293	0.0	0.0	0.6	0.2	0.0
r107	299	299	299.0	299	0.0	0.0	1.5	0.5	0.0
r108	308	308	308.0	308	0.0	0.0	2.4	0.9	0.1
r109	277	277	277.0	277	0.0	0.0	0.4	0.2	0.0
r110	284	284	284.0	284	0.0	0.0	3.8	1.3	0.0
r111	297	297	297.0	297	0.0	0.0	50.3	10.9	0.4
r112	298	298	298.0	298	0.0	0.0	10.6	3.3	0.0
rc101	219	219	219.0	219	0.0	0.0	0.5	0.2	0.0
rc102	266	266	266.0	266	0.0	0.0	1.7	0.4	0.0
rc103	266	266	266.0	266	0.0	0.0	9.6	2.0	0.1
rc104	*301*	301	301.0	301	0.0	0.0	0.7	0.3	0.2
rc105	244	244	244.0	244	0.0	0.0	10.0	4.3	0.2
rc106	252	252	252.0	252	0.0	0.0	1.0	0.3	0.0
rc107	277	277	277.0	277	0.0	0.0	0.9	0.3	0.0
rc108	298	298	298.0	298	0.0	0.0	0.2	0.1	0.0

Instance	BK	ILS			BG(%)	AG(%)	CPU(seconds)		
		Max	Avg	Min			Max	Avg	Min
c201	870	870	870.0	870	0.0	0.0	157.8	36.7	1.6
c202	930	930	930.0	930	0.0	0.0	185.0	59.0	20.8
c203	960	960	960.0	960	0.0	0.0	247.9	137.2	20.8
c204	980	980	974.0	970	0.0	0.6	246.5	217.6	104.9
c205	910	910	908.0	900	0.0	0.2	249.3	56.2	11.8
c206	930	930	927.0	920	0.0	0.3	219.8	111.5	5.8
c207	930	930	930.0	930	0.0	0.0	141.5	68.1	12.5
c208	950	950	950.0	950	0.0	0.0	68.3	33.3	4.9
r201	797	794	788.7	784	0.4	1.0	243.4	133.7	34.4
r202	930	921	910.3	896	1.0	2.1	269.8	165.6	25.4
r203	1021	**1026**	1011.3	996	-0.5	1.0	268.7	213.5	148.1
r204	1086	**1093**	1082.8	1071	-0.6	0.3	266.9	171.0	56.6
r205	953	953	948.4	942	0.0	0.5	253.2	169.9	28.8
r206	1029	1022	1012.4	1002	0.7	1.6	246.4	126.5	34.2
r207	1072	1067	1059.5	1049	0.5	1.2	248.3	174.0	77.1
r208	1112	**1113**	1107.6	1100	-0.1	0.4	267.9	165.6	47.6
r209	950	**956**	949.7	938	-0.6	0.0	231.6	145.8	76.3
r210	987	978	970.8	962	0.9	1.6	263.0	171.8	14.5
r211	1046	**1049**	1040.4	1025	-0.3	0.5	256.0	145.7	3.3
rc201	795	795	795.0	795	0.0	0.0	132.1	63.5	9.0
rc202	936	**938**	929.0	918	-0.2	0.7	262.1	156.2	31.1
rc203	1003	999	989.8	969	0.4	1.3	243.9	111.5	27.2
rc204	1140	1136	1131.3	1128	0.4	0.8	263.4	165.0	16.4
rc205	859	859	854.7	849	0.0	0.5	219.0	100.5	11.4
rc206	895	**899**	894.1	883	-0.4	0.1	255.9	152.0	52.1
rc207	983	983	952.1	941	0.0	3.1	252.9	129.9	14.7
rc208	1053	**1057**	1040.7	1020	-0.4	1.2	158.9	85.6	15.0

Table 4. Overall "Average" Comparison of ILS to the state-of-the-art methods.

Instance Set	Numb	IterLS		ACS*		SSA		GVNS		I3CH		ILS	
		$\overline{BG}$(%)	$\overline{CPU}$	$\overline{AG}$(%)	$\overline{CPU}^{\ddagger}$	$\overline{BG}$(%)	$\overline{CPU}$	$\overline{AG}$(%)	$\overline{CPU}$	$\overline{BG}$(%)	$\overline{CPU}$	$\overline{AG}$(%)	$\overline{CPU}^{\ddagger}$
c100	9	1.11	0.2	0.00	1.4	0.00	11.1	1.22	36.8	0.00	16.8	0.00	1.0
r100	12	1.90	0.1	0.24	84.8	0.11	12.3	2.68	6.5	0.56	19.1	0.00	1.8
rc100	8	2.92	0.1	0.00	31.7	0.00	11.7	3.51	2.2	1.66	17.0	0.00	1.0
c200	8	2.28	0.9	0.58	75.8	0.13	19.8	1.11	42.6	0.40	56.3	0.14	90.0
r200	11	2.90	0.9	3.17	344.4	1.30	24.1	3.38	7.5	1.05	117.4	0.93	162.1
rc200	8	3.43	0.9	2.04	341.7	0.96	26.5	3.96	3.5	2.68	79.6	0.97	120.5
pr01-10	10	4.74	0.9	1.22	359.8	0.98	59.1	1.62	2.7	1.07	72.7	0.74	50.4
pr11-20	10	9.56	1.0	11.87	196.4	3.71	85.6	4.26	5.4	4.28	86.8	2.12	97.9
Grand mean		3.64	0.6	2.49	183.9	0.94	31.9	2.73	12.6	1.44	59.1	0.63	65.6

$\ddagger$ Average computation time to obtain the best found

found BK (in seconds) from all runs since the experiments were based on the computational time. On the other hand, IterLS, SSA, GVNS and I3CH report the average of the computational time for solving each instance (in seconds). The computational time of all approaches were adjusted according the their computer's speed as summarized in Table 1.

From Table 4, we observe that ILS can produce better solutions against those of ACS* and GVNS. The $\overline{AG}$'s grand mean of ILS is only 0.63 %, whereas those of ACS* and GVNS are 2.49 % and 0.94 %, respectively. In terms of the computational time required to obtain the best found, ILS is much faster than ACS*. ILS only requires 65.6 s while ACS* requires 183.9 s, on average. Comparing against SSA, ILS finds better solutions at the expense of more computational time. IterLS is the fastest algorithm but the reported grand mean of $\overline{BG}$ is the largest.

Table 5 reports the best solutions obtained for each algorithm. ILS is the best compared against other methods. The grand mean of $\overline{BG}$ is only 0.23 %. The $\overline{BG}$s of ILS ranges from -0.04 % to 1.33 %, while the ones of IterLS and I3CH have wider ranges from 1.11 % to 9.56 % and 0.00 % to 4.28 %, respectively. ILS

Table 5. Overall "Best" Comparison of ILS to the state-of-the-art methods.

Instance set	Numb	IterLS	ACS*	SSA	GVNS	I3CH	ILS
		$\overline{BG}$(%)	$\overline{BG}$(%)	$\overline{BG}$(%)	$\overline{BG}$(%)	$\overline{BG}$(%)	$\overline{BG}$(%)
c100	9	1.11	0.00	0.00	0.56	0.00	0.00
r100	12	1.90	0.00	0.11	1.72	0.56	0.00
rc100	8	2.92	0.00	0.00	1.88	1.66	0.00
c200	8	2.28	0.40	0.13	0.55	0.40	0.00
r200	11	2.90	2.19	1.30	2.45	1.05	0.11
rc200	8	3.43	1.23	0.96	2.53	2.68	-0.04
pr01–10	10	4.74	1.06	0.98	0.56	1.07	0.34
pr11–20	10	9.56	11.13	3.71	3.17	4.28	1.33
Grand mean		3.64	2.09	0.94	1.71	1.44	0.23

Table 6. Comparison with the same computational time.

Instance set	$Numb$	I3CH	ILS		$\overline{CPU}$(s)
		$\overline{BG}$(%)	$\overline{BG}$(%)	$\overline{AG}$(%)	
c100	9	0.00	0.00	0.00	16.8
r100	12	0.56	0.00	0.00	19.1
rc100	8	1.66	0.00	0.03	17.0
c200	8	0.40	0.00	0.29	56.3
r200	11	1.05	0.36	1.24	117.4
rc200	8	2.68	0.20	1.30	79.6
pr01–10	10	1.07	0.30	0.80	72.7
pr11–20	10	4.28	1.29	2.30	86.8
Grand mean		1.44	0.28	0.76	59.1

obtains best known solutions for all instances of the first four instances sets. For rc200 instance set, a negative value of $\overline{BG}$ represents the improvement of some BKs.

I3CH outperforms SSA when using the same computational time that had been adjusted by their computers' speed [13]. We also compare the performance of ILS againts I3CH. As shown in Table 6, we found that the $\overline{BG}$'s grand mean of ILS is 1.16 % better than that of I3CH. ILS is able to obtain 0.00 % of $\overline{BG}$ for 4 out of 8 instance sets. In terms of $\overline{AG}$, ILS can reduce the grand mean of $\overline{AG}$ by almost 50 % compared against the one of $\overline{BG}$ of I3CH.

5 Conclusion

In this paper, we study the Orienteering Problem with Time Windows (OPTW). An algorithm based on Iterated Local Search (ILS) is proposed to solve the problem. Computational results have shown that the proposed algorithm is an effective algorithm. The algorithm has been able to improve 8 best known solution values of benchmark instances.

Other various mechanisms of ILS could be investigated. For instance, using other construction heuristics and restarting the algorithm from a new initial solution. It would also be interesting to consider applying ILS to other variants of the OP, e.g. the Team Orienteering Problem with Time Windows (TOPTW), the Time Dependent Orienteering Problem (TDOP) and the Tourist Trip Design Problem (TTDP).

Acknowledgements. This research is supported by Singapore National Research Foundation under its International Research Centre @ Singapore Funding Initiative and administered by the IDM Programme Office, Media Development Authority (MDA).

References

1. Tsiligirides, T.: Heuristic methods applied to orienteering. J. Oper. Res. Soc. **35**(9), 797–809 (1984)
2. Vansteenwegen, P., Souffriau, W., Van Oudheusden, D.: The orienteering problem: a survey. Eur. J. Oper. Res. **209**(1), 1–10 (2011)
3. Labadie, N., Mansini, R., Melechovský, J., Calvo, R.W.: The team orienteering problem with time windows: an LP-based granular variable neighborhood search. Eur. J. Oper. Res. **220**(1), 15–27 (2012)
4. Vansteenwegen, P., Souffriau, W., Vanden Berghe, G., Van Oudheusden, D.: Iterated local search for the team orienteering problem with time windows. Comput. Operat. Res. **36**(12), 3281–3290 (2009)
5. Goldberg, D.E.: Genetic Algorithms in Search, Optimization, and Machine Learning. Addison-Wesley, Reading (1989)
6. Kantor, M.G., Rosenwein, M.B.: The orienteering problem with time windows. J. Oper. Res. Soc. **43**(6), 629–635 (1992)
7. Righini, G., Salani, M.: Decremental state space relaxation strategies and initialization heuristics for solving the orienteering problem with time windows with dynamic programming. Comput. Oper. Res. **36**(4), 1191–1203 (2009)
8. Boland, N., Dethridge, J., Dumitrescu, I.: Accelerated label setting algorithms for the elementary resource constrained shortest path. Oper. Res. Lett. **34**(1), 58–68 (2006)
9. Montemanni, R., Gambardella, L.M.: Ant colony system for team orienteering problem with time windows. Found. Comput. Decis. Sci. **34**(4), 287–306 (2009)
10. Labadie, N., Mansini, R., Melechovský, J., Calvo, R.W.: Hybridized evolutionary local search algorithm for the team orienteering problem with time windows. J. Heuristics **17**(6), 729–753 (2011)
11. Montemanni, R., Weyland, D., Gambardella, L.M.: An enhanced ant colony system for the team orienteering problem with time windows. In: Proceedings of 2011 International Symposium on Computer Science and Society (ISCCS), pp. 381–384 (2011)
12. Lin, S.W., Yu, V.F.: A simulated annealing heuristic for the team orienteering problem with time windows. Eur. J. Oper. Res. **217**(1), 94–107 (2012)
13. Hu, Q., Lim, A.: An iterative three-component heuristic for the team orienteering problem with time windows. Eur. J. Oper. Res. **232**(2), 276–286 (2014)
14. Solomon, M.: Algorithms for the vehicle routing and scheduling problems with time window constraints. Oper. Res. **35**(2), 254–265 (1987)
15. Cordeau, J.F., Grendreau, M., Laporte, G.: A tabu search heuristic for periodic and multi-depot vehicle routing problems. Networks **30**(2), 105–119 (1997)

Analysis of Solution Quality of a Multiobjective Optimization-Based Evolutionary Algorithm for Knapsack Problem

Jun He[1]([✉]), Yong Wang[2], and Yuren Zhou[3]

[1] Department of Computer Science, Aberystwyth University, Aberystwyth, UK
jqh@aber.ac.uk
[2] School of Information Science and Engineering, Central South University,
Changsha 410083, China
[3] School of Advanced Computing, Sun Yat-sen University,
Guangzhou 510006, China

Abstract. Multi-objective optimisation is regarded as one of the most promising ways for dealing with constrained optimisation problems in evolutionary optimisation. This paper presents a theoretical investigation of a multi-objective optimisation evolutionary algorithm for solving the 0-1 knapsack problem. Two initialisation methods are considered in the algorithm: local search initialisation and greedy search initialisation. Then the solution quality of the algorithm is analysed in terms of the approximation ratio.

Keywords: Evolutionary algorithm · Knapsack problem · Multi-objective optimisation · Solution quality · Approximation ratio

1 Introduction

Consider the problem of maximizing an objective function,

$$\max_{\boldsymbol{x}} f(\boldsymbol{x}), \quad \text{subject to } g(x) \le 0. \tag{1}$$

The above constrained optimisation problem can be transferred into an unconstrained bi-objective optimisation problem. That is to optimize the original objective function plus to minimize the constraint violation simultaneously:

$$\begin{cases} \max_{\boldsymbol{x}} f(\boldsymbol{x}), \\ \min_{\boldsymbol{x}} v(\boldsymbol{x}), \end{cases} \tag{2}$$

where $v(\boldsymbol{x})$ is the degree of constraint violation, given by

$$v(\boldsymbol{x}) = \begin{cases} 0, & \text{if } g(\boldsymbol{x}) \le 0, \\ g(\boldsymbol{x}), & \text{otherwsie.} \end{cases} \tag{3}$$

The use of multi-objectives for single-objective optimisation problems could be traced back to 1990s [1]. This methodology has been termed multiobjectivisation [2]. Using multiobjectivization sometimes may help the search more efficient as shown in [3–6].

© Springer International Publishing Switzerland 2015
G. Ochoa and F. Chicano (Eds.): EvoCOP 2015, LNCS 9026, pp. 74–85, 2015.
DOI: 10.1007/978-3-319-16468-7_7

According to the survey [7], multi-objective optimisation is regarded as one of the most promising ways for dealing with constrained optimisation problems in evolutionary optimisation. A constrained optimisation problem is often transformed into a bi-objective optimisation problem, in which the first objective is the original objective function and the second objective is the degree of constraint violation [8–11]. After this transformation, Pareto dominance is frequently employed to compare individuals. Currently the research in this area is very active [7]. For example, a self-adaptive selection method is proposed recently in [12], which aims to exploit both non-dominated solutions with low constraint violations and feasible solutions with low objective function values. Multi-objective optimisation is combined with differential evolution in [13] and an infeasible solution replacement mechanism is proposed. A dynamic hybrid framework is presented in [14], where the global and local search models are implemented dynamically according to the feasibility proportion of the population.

This paper aims at analysing the solution quality of evolutionary algorithms (EAs) in terms of the approximation ratio. It is not intended to demonstrate that EAs are able to compete with problem-specific approximation algorithms, since this is unlikely in most cases. Nevertheless, it is still necessary and important to understand the solution quality of EAs, so the EAs with arbitrarily bad solution quality could be avoided in applications. The analysis of the approximation performance of EAs has attracted a lot of interests in recent years [15–17].

This paper investigate an existing multiobjective optimization-based EA [9] (MOEA) for solving constrained optimisation problems. The MOEA originally is designed for continuous optimization. Here it is adapted for solving the 0-1 knapsack problem. Although experiment results show its performance is good, no theoretical analysis exists for this MOEA [9]. This motivates our rigorous analysis.

The remainder of the paper is organized as follows. The 0-1 knapsack problem and approximation ratio are introduced in Sect. 2. The MOEA with the local search initialisation is analysed in Sect. 3. Section 4 is devoted to the analysis of the MOEA with the greedy search initialisation. Section 5 concludes the article.

2   0-1 Knapsack Problem and Approximation Ratio of Solutions

Given an instance of the 0-1 knapsack problem with a set of weights w_i, values v_i, and capacity W of a knapsack, the task is to find a binary string $x_{\max}$ so as to maximize the objective function,

$$\max_{x} f(x) = \sum_{i=1}^{n} v_i x_i, \quad \text{subject to } \sum_{i=1}^{n} w_i x_i \leq W, \tag{4}$$

where $x = (x_1 \cdots x_n)$ is a binary string. $x_i = 1$ if item i is selected in the knapsack; otherwise $x_i = 0$.

A *feasible solution* is a knapsack represented by an x which satisfies the constraint, that is $\sum_{i=1}^{n} w_i x_i \leq W$. An *infeasible* one is an x that violates the constraint. The string $(0 \ldots 0)$ represents a null knapsack. Without loss of generality, assume that a feasible solution always exists and n is large.

There exist well-known approximation algorithms for the 0-1 knapsack problem [18,19]. Probably the simplest one is the greedy search [18] whose worst-case approximation performance ratio equals to $1/2$ and time complexity is $O(n)$ plus $O(n \log n)$ for the initial sorting. A polynomial-time approximation scheme has been introduced in [18] whose worse-case performance is $k/(1+k)$ given an integer parameter k and its time complexity is $O(n^{k+1})$. Furthermore, a fully-polynomial-time approximation scheme is well-known [18] whose time complexity is $O(n/\epsilon^2)$ plus $O(n \log n)$ for the initial sorting given a parameter $\epsilon > 0$.

In evolutionary optimisation, the 0-1 knapsack problem has been taken as a benchmark in computer experiments [20,21] for evaluating the performance of various constraint-handling techniques. It is also one of favourite problems used in the theoretical study of EAs [22,23].

In order to assess the solution quality of an EA, an evolutionary approximation algorithm is defined as below. It follows the definition of conventional α-approximation algorithms [24, Definition 1].

Definition 1. *An EA is an α-approximation algorithm for a constrained optimisation problem if for all instances of the problem, the EA can produce a feasible solution in polynomial running time, whose objective function value is within a factor of α of that of an optimal solution. The* running time *of an EA is the expected number of function evaluations.*

In a maximisation problem (assume $f(x_{\max}) > 0$), a feasible solution x is called to have an α-*approximation ratio* if it satisfies

$$\frac{f(x)}{f(x_{\max})} \geq \alpha. \tag{5}$$

In order to prove that an EA is not an α-approximation algorithm, it is sufficient to show that an EA needs exponential running time to obtain a feasible solution with an α-approximation ratio in one instance of the problem.

3   Analysis of MOEA with Local Search Initialisation

The 0-1 knapsack problem can be transformed into a bi-objective optimisation problem, that is to maximize the objective function $f(x)$ and to minimize the constrain violation $v(x)$, where $v(\mathbf{x})$ is defined by

$$v(\mathbf{x}) = \begin{cases} 0; & \text{if } \mathbf{x} \text{ is feasible,} \\ \sum_{i=1}^{n} w_i - W, & \text{otherwise.} \end{cases} \tag{6}$$

Although a constrained optimisation problems can be converted into a bi-objective optimisation problem, there exists an essential difference between it and general multi-objective optimisation problems [9]. The target of general multi-objective optimisation is to obtain a final population with a diverse non-dominated individuals uniformly distributed on the Pareto front. However in the bi-objective optimisation problem derived from contained optimisation, the target is to obtain the optimal feasible solution of the original constrained optimisation problem. Consequently, there is no need to care about the uniform distribution of the resulting solutions on the Pareto front.

The MOEA adopted in this section is a variant of an existing MOEA proposed in [9]. The fundamental idea in this MOEA is that non-dominated individuals in a children population are chosen and replace dominated individuals of the parent population. The algorithm, based on Model 1 of [9], is described in Algorithm 1 for solving the 0-1 knapsack problem.

Algorithm 1. MOEA [9]

1: initialize population Φ_0;
2: **for** $t = 0, 1, 2, \cdots$ **do**
3:    perform bitwise mutation and generate a children population $\Phi_{t.a}$ with N individuals;
4:    evaluate the values of $f(\boldsymbol{x})$ and $v(\boldsymbol{x})$;
5:    choose the non-dominated individuals from population $\Phi_{t.a}$ and assume there are k non-dominated individuals, denoted as $\{\boldsymbol{x}_1, \cdots, \boldsymbol{x}_k\}$;
6:    set an intermediate population $\Phi_{t.b} \leftarrow \Phi_t$;
7:    **for** $i = 1, \cdots, k$ **do**
8:       let m be the number of individuals in $\Phi_{t.b}$ which are dominated by $\boldsymbol{x}_i$;
9:       **if** $m = 0$ **then**
10:          do nothing;
11:       **else if** $m = 1$ **then**
12:          the corresponding dominated individual is replaced by $\boldsymbol{x}_i$;
13:       **else**
14:          **if** the dominated individuals are feasible **then**
15:             the individual with the smallest objective function value is replaced by $\boldsymbol{x}_i$;
16:          **else**
17:             one of the dominated individuals is randomly chosen and replaced by $\boldsymbol{x}_i$;
18:          **end if**
19:       **end if**
20:    **end for**
21:    set the next generation population $\Phi_{t+1} \leftarrow \Phi_{t.b}$.
22: **end for**

It should be pointed out that the running time of EAs is dependent on initialisation. Two initialisation methods are considered in the paper: initialisation by the local search and by the greedy search. In this section, we investigate the first one: the local search initialisation, described in Algorithm 2. This initialisation does not only produce both feasible solutions (local optima), but also infeasible solutions. Bitwise mutation flips each bit of a binary string with probability $\frac{1}{n}$.

Population size N is set to a large constant and for the sake of analysis, assume that $N/4$ is an integer.

Algorithm 2. Local Search Initialisation

1: set $x = (0 \cdots 0)$;
2: **while** x is feasible **do**
3:    flip one 0-valued bit of x into 1-valued, denote it by y;
4:    **if** y is feasible **then**
5:       let $x \leftarrow y$;
6:    **else**
7:       let $x \leftarrow y$ with probability $1/2$;
8:    **end if**
9: **end while**
10: repeat the above steps until N individuals are produced.

We show that the solution quality of the MOEA with the local search initialisation might be arbitrarily bad using the following instance of the 0-1 knapsack problem (Table 1).

Instance 1. *In the following table, H, I and J represent index sets. For the sake of simplicity, assume that $\frac{n}{2}$ and $\frac{\alpha n}{2}$ are integers. Fixing a constant $\alpha \in (0, 1)$, choose n an enough large integer so that $n > \frac{2}{\alpha}$ and $\frac{2}{\alpha} > \frac{n\alpha^2 \ln n}{2}$.*

Table 1. Instance 1

	H	I	J
i	1	$2, \cdots, \frac{n}{2} + 1$	$\frac{n}{2} + 2, \cdots, n$
v_i	n	1	$\alpha^{\ln n}$
w_i	n	$\frac{2}{\alpha}$	$\alpha^{2 \ln n}$
W	n		

Let $x_{\max}$ represent the global optimum such that $x_1 = 1$ and other bits $x_i = 0$. The global optimum is unique. Its objective function value is

$$f(x_{\max}) = n. \tag{7}$$

Let x_{loc} represent a local optimum[1] such that $\frac{\alpha n}{2}$ bits $x_i = 1$ (where $i \in I$) and other bits $x_i = 0$. Its objective function value is

$$f(x_{\mathrm{loc}}) = \frac{\alpha n}{2}. \tag{8}$$

---

[1] A feasible solution x is called a *local optimum* if $f(y) < f(x)$ for any feasible solution y within Hamming distance $d(x, y) = 1$.

$f(x_{\text{loc}})$ is the second largest objective function value among feasible solutions. The number of such local optima x_{loc} is experiential in n,

$$\binom{\frac{n}{2}}{\frac{\alpha n}{2}} \geq \left(\frac{1}{\alpha}\right)^{\frac{\alpha n}{2}}. \tag{9}$$

Let x_{vio1} denote an infeasible solution such that $x_i = 1$ for $\frac{\alpha n}{2}$ indexes $i \in I$, one $i \in J$, and $x_i = 0$ for other i. Its objective function value and violation value are

$$f(x_{\text{vio1}}) = f(x_{\text{loc}}) + \alpha^{\ln n}, \quad v(x_{\text{vio1}}) = \alpha^{2\ln n}. \tag{10}$$

The number of such infeasible solutions x_{vio1} is exponential in n,

$$\binom{\frac{n}{2}}{\frac{\alpha n}{2}} \frac{n-1}{2} \geq \frac{n-1}{2}\left(\frac{1}{\alpha}\right)^{\frac{\alpha n}{2}}. \tag{11}$$

Let x_{vio2} denote an infeasible solution such that $x_1 = 1$, $x_i = 1$ for one $i \in J$, and $x_i = 0$ for any other i. Its objective function value and violation value are

$$f(x_{\text{vio2}}) = f(x_{\text{max}}) + \alpha^{\ln n}, \quad v(x_{\text{vio2}}) = \alpha^{2\ln n}. \tag{12}$$

It is easy to verify that the degree of constraint violation $v(x_{\text{vio1}})$ and $v(x_{\text{vio2}})$ $(= \alpha^{\ln n})$ is the minimum among all infeasible solutions.

Instance 1 is hard since Hamming distance between a local optimum x_{loc} and the unique global optimum x_{max} is large $\geq \alpha n/2$ and the the number of local optima is exponential in n. Since the selection used in the MOEA is that non-dominated individuals in a children population are chosen and replace dominated individuals of the parent population, it prevents individuals moving from the 2nd best fitness level to the best fitness level. Thus the EA needs exponential time to leave the absorbing basin of the local optima.

Theorem 1. *For Instance 1 and any constant $\alpha \in (0,1)$, the MOEA with the local search initialisation needs $\Omega(n^{\frac{\alpha n}{2}})$ running time to find an α-approximation solution in the worst case.*

Proof. After the initialisation, individuals generated by the local search may include x_{max}, local optima x_{loc} and infeasible solutions x with Hamming distance $H(x, x_{\text{max}}) = 1$ or $H(x, x_{\text{loc}}) = 1$. The worst case is that after initialisation, population Φ_0 is composed of N local optima x_{loc} and infeasible solutions x_{vio1}. Notice that the number of local optima x_{loc} and infeasible solutions x_{vio1} is exponential in n. The individuals in Φ_0 may be chosen to be different.

Assume that in the tth generation, population Φ_t is composed of N local optima x_{loc} and infeasible solutions x_{vio1}. The approximation ratio between $f(x_{\text{loc}})$ and $f(x_{\text{max}})$ is

$$\frac{f(x_{\text{loc}})}{f(x_{\text{max}})} = \frac{\alpha}{2} < \alpha. \tag{13}$$

Since α could be any constant, the above approximation ratio could arbitrarily bad.

As we know that x_{max} is the unique solution satisfying $f(x_{max}) > f(x_{loc})$, it is sufficient to prove that the EA needs exponential running time to generate x_{max}.

First we consider the event of mutating x_{loc} or x_{vio1} into a child y. The event can be decomposed into the following mutually exclusive and exhaustive sub-events.

1. y is a feasible solution such that $f(y) < f(x_{loc})$.
   Obviously y will not dominate x_{loc}. At the same time, y will not dominate x_{vio1} since $f(y) < f(x_{loc}) < f(x_{vio1})$. Thus y will not dominate any individuals in Φ_t and cannot be selected into the next generation population.
2. y is a feasible solution such that $f(y) = f(x_{loc})$, that is, y is a x_{loc}.
3. y is a feasible solution such that $f(y) > f(x_{loc})$, that is, y is the global optimum x_{max}.
   In this case, all 1-valued bits x_i (where $i \in I$) must be flipped from 1 to 0. The probability of the event happening is at most

$$O \left(\frac{1}{n} \right)^{\frac{\alpha n}{2}} . \tag{14}$$

4. y is an infeasible solution such that $v(y) > v(x_{vio1})$.
   In this case, y will not dominate x_{loc} and x_{vio1}, so it cannot be selected into the next generation population.
5. y is an infeasible solution such that $v(y) = v(x_{vio1})$, that implies,
   (a) either y is x_{vio1};
   (b) or y is an infeasible solution x_{vio2}.
      In the second case, all 1-valued bits x_i (where $i \in I$) must be flipped from 1 to 0. The probability of the event happening is

$$O \left(\frac{1}{n} \right)^{\frac{\alpha n}{2}} . \tag{15}$$

From the above analysis, we observe that only when either x_{max} or an infeasible solution x_{vio2} is generated via mutation, the population status could be changed. Otherwise the population is still composed of N solutions x_{loc} and x_{vio1}. The probability of the former event happening is $O \left(\frac{1}{n} \right)^{\frac{\alpha n}{2}}$.

Next we analyse the role of using a population. Consider the event that a population includes either x_{max} or an infeasible solution x_{vio2} is generated via mutation. Since N parents are selected and mutated independently, the probability of the event happening is $N O \left(n^{\frac{2}{\alpha n}} \right)$. Thus the expected number of generations for the EA to reach x_{max} is $\frac{1}{N} \Omega \left(n^{\frac{\alpha n}{2}} \right)$. Since there are N fitness evaluations at each generation, the expected number of fitness evaluations is $\Omega \left(n^{\frac{\alpha n}{2}} \right)$. The required conclusion is then proven. □

4   Analysis of MOEA with Greedy Search Initialisation

In order to produce a good quality solution with a guaranteed approximation ratio, a natural idea is to combine an EA with an approximation algorithm: first we apply an approximation algorithm to producing approximation solutions as the initial population, and then apply the MOEA to searching a global optimum. In this section, we consider a $\frac{1}{2}$-approximation algorithm to implement the initialisation. It a variant of the greedy search [18, Sect. 2.4], described in Algorithm 3. Notice that the initialisation does not only produce feasible solutions (local optima), but also infeasible solutions.

Algorithm 3. Greedy Search Initialisation

1: sort all the items via their values so that $p_1 \geq \cdots \geq p_n$;
2: then greedily add the items in the above order to the knapsack as long as adding an item to the knapsack does not exceeding the capacity of the knapsack. Denote the solution by x_a;
3: resort all the items via the ratio of their values to their corresponding weights so that $\frac{p_1}{w_1} \geq \cdots \geq \frac{p_n}{w_n}$;
4: Then greedily add the items in the above order to the knapsack as long as adding an item to the knapsack does not exceeding the capacity of the knapsack. Denote the solution by x_b;
5: put x_a and x_b into the initial population;
6: repeat the above procedure until $\frac{N}{2}$ individuals are produced;
7: for each of these $\frac{N}{2}$ individuals, add one item and then $\frac{N}{2}$ infeasible solutions are produced.

Using the greedy search initialisation, we may find the global optimal solution of Instance 1 during the initialisation phase. Furthermore, since the greedy search is a 1/2-approximation algorithm for the 0-1 knapsack problem, the MOEA with the greedy search initialisation is an evolutionary 1/2-approximation algorithm too. The advantage of using an EA is the ability to obtain the global optimum due to the use of bitwise mutation.

In the following we answer the question: can the MOEA with the greedy search initialisation find a solution with the approximation ratio better than 1/2? Through analysing the instance described below, we obtain a negative answer (Table 2).

Instance 2. *In the following table, H, I, J and K represent index sets. For the sake of simplicity, assume that n/4 is an integer.*

Let $x_{\max}$ represent the unique global optimum such that $x_i = 1$ for any $i \in H$ and $x_i = 0$ for any other i. Its objective function value is

$$f(x_{\max}) = 2n. \tag{16}$$

Table 2. Instance 2

	H	I	J	K
i	$1,2$	$3,\cdots,\frac{n}{4}$	$\frac{n}{4}+1,\cdots,\frac{n}{2}$	$\frac{n}{2}+1,\cdots,n$
v_i	n	$n+2$	n^{-3}	n^{-3}
w_i	n	$n+1$	n^{-4}	$\frac{1}{4(1+n^{-4})}$
W	$2n$			

Let $\boldsymbol{x}_{\mathrm{loc}}$ represent a local optimum such that $x_i = 1$ for one $i \in I$, any $i \in J$ and $\frac{n}{4} - 2$ indexes $i \in K$; $x_i = 0$ for all other i. Its objective function value is

$$f(\boldsymbol{x}_{\mathrm{loc}}) = n + 2 + \left(\frac{n}{2} - 2\right) n^{-3}. \tag{17}$$

$f(\boldsymbol{x}_{\mathrm{loc}})$ is the second largest objective function value among feasible solutions.

Let $\boldsymbol{x}_{\mathrm{vio1}}$ represent an infeasible solution such that $x_i = 1$ for one $i \in I$, any $i \in J$ and $\frac{n}{4} - 1$ indexes $i \in K$; $x_i = 0$ for all other i. Its objective function value and violation value are

$$f(\boldsymbol{x}_{\mathrm{vio1}}) = f(\boldsymbol{x}_{\mathrm{loc}}) + n^{-3}, \quad v(\boldsymbol{x}_{\mathrm{vio1}}) = \frac{1}{4(1 + n^{-4})}. \tag{18}$$

Let $\boldsymbol{x}_{\mathrm{vio2}}$ represent an infeasible solution such that $x_i = 1$ for any $i \in H$ and one $i \in K$, and $x_i = 0$ for any other i. Its objective function value and violation value satisfy

$$f(\boldsymbol{x}_{\mathrm{vio2}}) = f(\boldsymbol{x}_{\mathrm{max}}) + n^{-3}, \quad v(\boldsymbol{x}_{\mathrm{vio2}}) = \frac{1}{4(1 + n^{-4})}. \tag{19}$$

Let $\boldsymbol{x}_{\mathrm{vio3}}$ represent an infeasible solution such that $x_i = 1$ for any $i \in H$ and at least one $i \in J$, and $x_i = 0$ for any other i. Its objective function value and violation value satisfy

$$f(\boldsymbol{x}_{\mathrm{max}}) < f(\boldsymbol{x}_{\mathrm{vio3}}) \le f(\boldsymbol{x}_{\mathrm{max}}) + \frac{n^{-2}}{4}, \quad 0 < v(\boldsymbol{x}_{\mathrm{vio3}}) < \frac{n^{-3}}{4}. \tag{20}$$

Instance 2 is a hard problem to EAs using bitwise mutation since Hamming distance between a local optimum $\boldsymbol{x}_{\mathrm{loc}}$ and the unique global optimum $\boldsymbol{x}_{\mathrm{max}}$ is large. Another trouble is the number of local optima which is exponential in n. Thus it is difficult for a population to leave the absorbing basin of the local optima.

Theorem 2. *For Instance 2, the MOEA with the greedy search initialisation can find a $\frac{1}{2}$-approximation solution after initialisation. But in the worst case, it needs $\Omega(n^{\frac{n}{4}})$ running time to find a $(\frac{1}{2} + \frac{1}{n} + \frac{1}{2n^3})$-approximation solution.*

Proof. The first conclusion is trivial due to the use of the greedy search. After the initialisation, individuals generated by the greedy search are local optima

x_{loc} and infeasible solutions x with Hamming distance $H(x, x_{loc}) = 1$. The local optimum x_{loc} has an approximation ratio given by

$$\frac{f(x_{loc})}{f(x_{max})} = \frac{n + 2 + \left(\frac{n}{2} - 2\right) n^{-3}}{2n} \in \left(\frac{1}{2}, \frac{1}{2} + \frac{1}{n} + \frac{1}{2n^3}\right). \qquad (21)$$

The proof of the second conclusion is similar to that of Theorem 1. The worst case is that after initialisation, population Φ_0 is composed of N local optima x_{loc} and infeasible solutions x_{viol}. Notice that the number of local optima x_{loc} and infeasible solutions x_{viol} is exponential in n. The individuals in Φ_0 could be chosen to be different.

Assume that in the tth generation, population Φ_t is composed of N different local optima x_{loc} and infeasible solutions x_{viol}. Let $\mathbf{y}$ be a child mutated from a parent $\mathbf{x}$. The event can be decomposed into the following mutually exclusive and exhaustive sub-events.

1. ($\mathbf{y}$ is feasible such that $f(\mathbf{y}) < f(\mathbf{x}_{loc})$.
   According to the selection based on the Pareto-dominance, $\mathbf{y}$ will not be selected to the next generation population.
2. $\mathbf{y}$ is feasible such that $f(\mathbf{y}) = f(\mathbf{x}_{loc})$, that is, $\mathbf{y}$ is a local optimum x_{loc}.
3. $\mathbf{y}$ is feasible such that $f(\mathbf{y}) > f(\mathbf{x}_{loc})$, that is, $\mathbf{y}$ is x_{max}.
   In this case, any 1-valued bit x_i where $i \in I \cup J \cup K$ must be flipped from 1 to 0, and x_1 and x_2 must be flipped from 0 to 1. Thus at least $\frac{n}{4}$ bits must be flipped. The probability of this event happening is at most $O(n^{-\frac{n}{4}})$.
4. $\mathbf{y}$ is infeasible such that $v(\mathbf{x}) = \frac{1}{4(1+n^{-4})}$, that is
   – either $\mathbf{y}$ is x_{viol};
   – or $\mathbf{y}$ is x_{vio2}.
   In the second case, except one 1-valued bit x_i where $i \in K$, any other 1-valued bit x_i where $i \in I \cup J \cup K$ must be flipped from 1 to 0, and x_1 and x_2 must be flipped from 0 to 1. Thus at least $\frac{n}{4}$ bits must be flipped. The probability of this event happening is at most $O(n^{-\frac{n}{4}})$.
5. $\mathbf{y}$ is infeasible such that $v(\mathbf{x}) < \frac{1}{4(1+n^{-4})}$, that means, $\mathbf{y}$ is x_{vio3}.
   In this case, any 1-valued bit x_i where $i \in I \cup K$ must be flipped from 1 to 0, and x_1 and x_2 must be flipped from 0 to 1. Thus at least $\frac{n}{4}$ bits must be flipped. The probability of this event happening is at most $O(n^{-\frac{n}{4}})$.
6. $\mathbf{y}$ is infeasible such that $v(\mathbf{x}) > \frac{1}{4(1+n^{-4})}$. $\mathbf{y}$ will not be selected since it is dominated by x_{viol}.

From the above analysis, we observe that the probability of generating a non-x_{loc} and non-x_{viol} child and selecting it to the next generation population is small, that is $O(n^{-\frac{n}{4}})$.

Next we analyse the role of using a population. Consider the event that the next generation population includes a non-x_{loc} or non-x_{viol} child. Since N parents are mutated independently, the probability of the event happening is at most $NO(n^{-\frac{n}{4}})$. This implies that the expected number of generations for the EA to reach x_{max} is at least $\frac{1}{N}\Omega(n^{\frac{n}{4}})$. Since there are N fitness evaluations at each generation, the expected number of fitness evaluations is $\Omega(n^{\frac{n}{4}})$. The required conclusion is then proven. □

The above theorem shows that the MOEA with the greedy search initialisation finds a $(\frac{1}{2} + \frac{1}{n} + \frac{1}{2n^3})$-approximation solution in $\Omega(n^{\frac{n}{4}})$ running time. As $n \to +\infty$, the approximation ratio goes towards $1/2$. In other words, the MOEA doesn't substantially improve the solution quality since the greedy search already produces a $1/2$ approximation solution during initialisation.

5   Conclusions

This paper has assessed the solution quality of an existing MOEA [9] for solving the 0-1 knapsack problem. The solution quality of an EA is measured in terms of the approximation ratio. Two different initialization methods are analysed in the MOEA: local search initialisation and greedy search initialisation.

When the initial population is produced by the local search, the solution quality of the MOEA might be arbitrarily bad in some instance. That is, given any constant $\alpha \in (0, 1)$, the MOEA needs $\Omega(n^{\frac{\alpha n}{2}})$ running time to find an α-approximation solution in the worst case in some instance.

When the initial population is produced by the greedy search, the MOEA may guarantee a $1/2$-approximation solution within polynomial time. However, this improvement is caused by the use of the greedy search, rather than the MOEA itself. In some instance, the MOEA with the greedy search initialisation needs $\Omega(n^{\frac{n}{4}})$ running time to find a $(\frac{1}{2} + \frac{1}{n} + \frac{1}{2n^3})$-approximation solution. In other words, the MOEA doesn't substantially improve the solution quality comparing with the greedy search.

Other types of initialisation, such as random initialisation, are not considered in the current paper. It is left for future work.

Acknowledgement. This work was partially supported by EPSRC under Grant No. EP/I009809/1 (He), by NSFC under Grant No. 61170081, 61472143 (Zhou), 61273314 and by the Program for New Century Excellent Talents in University under Grant NCET-13-0596 (Wang).

References

1. Louis, S.J., Rawlins, G.: Pareto optimality, GA-easiness and deception. In: Proceedings of 5th International Conference on Genetic Algorithms, Morgan Kaufmann, pp. 118–123 (1993)
2. Knowles, J.D., Watson, R.A., Corne, D.W.: Reducing local optima in single-objective problems by multi-objectivization. In: Zitzler, E., Deb, K., Thiele, L., Coello Coello, C.A., Corne, D.W. (eds.) EMO 2001. LNCS, vol. 1993, p. 269. Springer, Heidelberg (2001)
3. Jensen, M.T.: Helper-objectives: Using multi-objective evolutionary algorithms for single-objective optimisation. J. Math. Model. Algorithms **3**(4), 323–347 (2005)
4. Neumann, F., Wegener, I.: Minimum spanning trees made easier via multi-objective optimization. Nat. Comput. **5**(3), 305–319 (2006)

5. Neumann, F.: Expected runtimes of a simple evolutionary algorithm for the multi-objective minimum spanning tree problem. Eur. J. Oper. Res. **181**(3), 1620–1629 (2007)

6. Friedrich, T., He, J., Hebbinghaus, N., Neumann, F., Witt, C.: Approximating covering problems by randomized search heuristics using multi-objective models. Evol. Comput. **18**(4), 617–633 (2010)

7. Segura, C., Coello, C.A.C., Miranda, G., León, C.: Using multi-objective evolutionary algorithms for single-objective optimization. 4OR **11**(3), 201–228 (2013)

8. Zhou, Y., Li, Y., He, J., Kang, L.: Multi-objective and MGG evolutionary algorithm for constrained optimisation. In: Proceedings of 2003 IEEE Congress on Evolutionary Computation, Canberra, Australia, pp. 1–5. IEEE Press (2003)

9. Cai, Z., Wang, Y.: A multiobjective optimization-based evolutionary algorithm for constrained optimization. IEEE Trans. Evol. Comput. **10**(6), 658–675 (2006)

10. Wang, Y., Cai, Z., Guo, G., Zhou, Y.: Multiobjective optimization and hybrid evolutionary algorithm to solve constrained optimization problems. IEEE Trans. Syst. Man Cybern. Part B **37**(3), 560–575 (2007)

11. Wang, Y., Cai, Z., Zhou, Y., Zeng, W.: An adaptive tradeoff model for constrained evolutionary optimization. IEEE Trans. Evol. Comput. **12**(1), 80–92 (2008)

12. Jiao, L., Li, L., Shang, R., Liu, F., Stolkin, R.: A novel selection evolutionary strategy for constrained optimization. Inf. Sci. **239**, 122–141 (2013)

13. Wang, Y., Cai, Z.: Combining multiobjective optimization with differential evolution to solve constrained optimization problems. IEEE Trans. Evol. Comput. **16**(1), 117–134 (2012)

14. Wang, Y., Cai, Z.: A dynamic hybrid framework for constrained evolutionary optimization. IEEE Trans. Syst. Man Cybern. Part B Cybern. **42**(1), 203–217 (2012)

15. Friedrich, T., Oliveto, P., Sudholt, D., Witt, C.: Analysis of diversity-preserving mechanisms for global exploration. Evol. Comput. **17**(4), 455–476 (2009)

16. Oliveto, P.S., He, J., Yao, X.: Analysis of the (1+1)-EA for finding approximate solutions to vertex cover problems. IEEE Trans. Evol. Comput. **13**(5), 1006–1029 (2009)

17. Lai, X., Zhou, Y., He, J., Zhang, J.: Performance analysis of evolutionary algorithms for the minimum label spanning tree problem. IEEE Trans. Evol. Comput. **18**(6), 860–872 (2014)

18. Martello, S., Toth, P.: Knapsack Problems: Algorithms and Computer Implementations. Wiley, New York (1990)

19. Kellerer, H., Pferschy, U., Pisinger, D.: Knapsack Problems. Springer, Heidelberg (2004)

20. Michalewicz, Z., Arabas, J.: Genetic algorithms for the 0/1 knapsack problem. In: Raś, Z.W., Zemankova, M. (eds.) ISMIS 1994. LNCS, vol. 869, pp. 134–143. Springer, Heidelberg (1994)

21. Michalewicz, Z.: Genetic Algorithms + Data Structures = Evolution Programs, 3rd edn. Springer, New York (1996)

22. Kumar, R., Banerjee, N.: Analysis of a multiobjective evolutionary algorithm on the 0-1 knapsack problem. Theor. Comput. Sci. **358**(1), 104–120 (2006)

23. Zhou, Y., He, J.: A runtime analysis of evolutionary algorithms for constrained optimization problems. IEEE Trans. Evol. Comput. **11**(5), 608–619 (2007)

24. Williamson, D.P., Shmoys, D.B.: The Design of Approximation Algorithms. Cambridge University Press, New York (2011)

Evolving Deep Recurrent Neural Networks Using Ant Colony Optimization

Travis Desell[1]($\boxtimes$), Sophine Clachar[1], James Higgins[2], and Brandon Wild[2]

[1] Department of Computer Science, University of North Dakota, Grand Forks, USA
tdesell@cs.und.edu, sophine.clachar@my.und.edu
[2] Department of Aviation, University of North Dakota, Grand Forks, USA
{jhiggins,bwild}@aero.und.edu

Abstract. This paper presents a novel strategy for using ant colony optimization (ACO) to evolve the structure of deep recurrent neural networks. While versions of ACO for continuous parameter optimization have been previously used to train the weights of neural networks, to the authors' knowledge they have not been used to actually design neural networks. The strategy presented is used to evolve deep neural networks with up to 5 hidden and 5 recurrent layers for the challenging task of predicting general aviation flight data, and is shown to provide improvements of 63 % for airspeed, a 97 % for altitude and 120 % for pitch over previously best published results, *while at the same time not requiring additional input neurons for residual values*. The strategy presented also has many benefits for neuro evolution, including the fact that it is easily parallizable and scalable, and can operate using any method for training neural networks. Further, the networks it evolves can typically be trained in fewer iterations than fully connected networks.

Keywords: Ant colony optimization · Time-series prediction · Neural networks · Flight prediction · Aviation informatics

1 Introduction

Neural networks have been widely used for time series data prediction [11,43]. Unfortunately, current popular techniques for designing and training neural networks such as convolutional and deep learning strategies, popular within computer vision, do not easily apply to time series prediction. This is in part because the number of input parameters is relatively small (compared to pixels within images), the fact they do not easily deal with recurrent memory neurons, and the goal is prediction, as opposed to classification. Even more problematic, these strategies do not help address the rather challenging problem of determining the best performing structure for those neural networks. Automated strategies for simultaneously evolving the structure and weights of neural networks have been examined through strategies such as NeuroEvolution of Augmenting Topologies (NEAT) [37] and Hyper-NEAT [38], and while these can evolve recurrent connections, they require non-trivial modification to evolve the recurrent memory nureons typically used for time series prediction.

© Springer International Publishing Switzerland 2015
G. Ochoa and F. Chicano (Eds.): EvoCOP 2015, LNCS 9026, pp. 86–98, 2015.
DOI: 10.1007/978-3-319-16468-7_8

Recent work in using neural networks for time series prediction has involved utilizing *residuals* or *lags* similar to the Auto-Regressive Integrated Moving Average (ARIMA) model [42], as done by Khashei *et al.* [23] and Omer *et al.* [30]. Other work has investigated strategies for cooperative co-evolution for Elman recurrent neural networks [9,10], however these strategies involve single parameter time series data such as the Mackey-Glass, Lorenz and Sunspot data sets.

Ant colony optimization (ACO) [6,17,19] is an optimization technique originally designed for use on discrete problems, with a common example being the Traveling Salesman Problem [18]. It has since been extended for use in continuous optimization problems [5,20,27,34–36], including training artificial neural networks [3,7,24,31,40]. While ACO has been studied for training artificial neural networks (ANNs), to the authors' knowledge there is little work in using ACO to actually *design* neural networks, with the closest being Sivagaminathan *et al.* using ACO to select input features for neural networks [33].

This work presents a novel strategy based on ant colony optimization which evolves the structure of recurrent deep neural networks with multiple input data parameters. While ant colony optimization is used to evolve the network structure, any number of optimization techniques can be used to optimize the weights of those neural networks. Trained neural networks with good fitness will be used to update the pheromones, reinforcing connections between neurons that provide good solutions. The algorithm is easily parallelizable and scalable, using a steady state population of best performing neural networks to determine when pheromones are incremented, and any number of worker processes can asynchronously train neural networks generated by the ant colony optimization strategy.

This algorithm is evaluated using the real world problem of predicting general aviation flight data, and compared to previously best published results for a set of testing data. For three of the four parameters evaluated (airspeed, altitude, and pitch), this approach improves significantly on previously published results, while at the same time not requiring additional input nodes for ARIMA residuals. For the fourth parameter, roll, the strategy performs worse, however this may be due to the fact that the neural networks were not trained for long enough. The authors feel that the results provide a strong case for the use of ant colony optimization in the design of neural networks, given its ability to find novel and effective neural network topologies that can be easily trained (apart from the roll parameter which requires further study).

2   Predicting General Aviation Flight Data

General aviation comprises 63 % of all civil aviation activity in the United States; covering operation of all non-scheduled and non-military aircraft [21,32]. While general aviation is a valuable and lucrative industry, it has the highest accident rates within civil aviation [29]. For many years, the general aviation accident and fatality rates have hovered around 7 and 1.3 per 100,000 flight hours, respectively [1]. The general aviation community and its aircraft are very diverse, limiting the utility of the traditional flight data monitoring (FDM) approach used by commercial airlines.

The National General Aviation Flight Information Database (NGAFID) has been developed at the University of North Dakota as a central repository for general aviation flight data. It consists of per-second flight data recorder (FDR) data from three fleets of aircraft. As of November 2014, the database stores FDR readings from over 200,000 flights, with more being added daily. It currently stores over 750 million per-second records of flight data. The NGAFID provides an invaluable source of information about general aviation flights, as most of these flights are from aviation students, where there is a wider variance in flight parameters than what may normally be expected within data from professionally piloted flights.

Having algorithms which can accurately predict FDR parameters would be able to not only warn pilots of problematic flight behavior, but also be used to predict impending failures of engines and other hardware. As such, investigating predictive strategies such as these has the potential to reduce costs for maintaining general aviation fleets, and more importantly save lives.

3   Previous Results

In previous work, the authors evaluated a suite of feed forward, Jordan and Elman recurrent neural networks to predict flight parameters [14]. This work was novel in that to our knowledge, neural networks have not been previously applied to predicting general aviation flight data. These results were encouraging in that some parameters such as altitude and airspeed can be predicted with high accuracy, at 0.22–0.62 % for airspeed, 0.026–0.08 % for airspeed, 0.88–1.49 % for pitch and 0.5–2 % for roll. These neural networks were trained using backpropagation via stochastic gradient descent, gradient descent from a baseline predictor (which mimicked how deep neural networks are currently trained by pre-training each layer to predict its input), and with asychronous differential evolution (ADE). ADE was shown to significantly outperform both types of backpropagation, provided solutions with up to 70 % improvement. It was also shown that while ADE outperformed backpropagation, it still had trouble training the larger fully connected Jordan and Elman recurrent neural networks (which provided the best predictions), motivating further study.

4   Methodology

The ACO based strategy works as follows. Given a potentially fully connected recurrent neural network – where each node has a potential connection to every node in the subsequent layer and to a respective node in the recurrent layer – each connection between neurons can be seen as a potential path for an ant (see Fig. 1). Every potential connection is initialized with a base amount of pheromone, and the master process stores the amount of pheromone on each connection. Worker processes receive neural network designs generated by taking a selected number of ants, and having them choose a path through the fully connected neural network biased by the amount of pheromone on each connection.

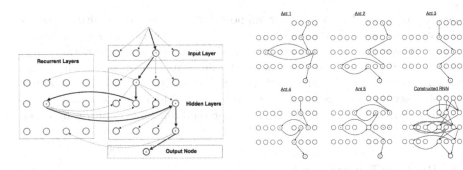

Fig. 1. Ants select a forward propagating path through neurons randomly based on the pheromone on each path, assuming a fully connected strategy between the input, hidden and output layers; and a potential connection from each hidden node to a respective node in the recurrent layer that is fully connected back to its hidden layer.

Fig. 2. The server creates neural networks for the workers to evaluate by combining the paths selected by a given number of ants. This generates various Elmann-like neural networks which have less training complexity than a fully connected Elman design.

Multiple ants can choose the connections between neurons. Those ant paths are be combined to construct a neural network design which is sent to worker processes and trained on the input flights using backpropagation, evolutionary algorithms or any other neural network training algorithm. The master process maintains a population of the best neural network designs, and when a worker reports the accuracy of a newly trained neural network, if it improves the population, the master process will increase the pheromone on every connection that was in that neural network. The master process periodically degrades pheromone levels, as is done in the standard ACO algorithm. This strategy allows the evolution of recurrent neural networks with potentially many hidden layers and hidden nodes, to determine what design can best predict flight parameters (Fig. 2).

5   Results

5.1   Optimization Software, Data and Reproducibility

Given the complexity of examining complex neural networks over per-second flight data, a package requiring easy use of high performance computing resources was required. While there exist some standardized evolutionary algorithms packages [2,8,25,41], as well as those found in the R programming language [4,28] and MATLAB [26], they do not easily lend themselves towards use in high performance computing environments.

This work utilizes the Toolkit for Asynchronous Optimization (TAO), which is used by the MilkyWay@Home volunteer computing to perform massively distributed evolutionary algorithms on tens of thousands of volunteered hosts [12,15,16]. It is implemented in C++ and MPI, allowing easy use on clusters and supercomputers, and also provides support for systems with multiple graphical processing units. Further, TAO has shown that performing evolutionary algorithms asynchronously can provide significant improvements to performance and scalability over iterative approaches [13,39]. TAO is open source and freely available on GitHub, allowing easy use and extensibility[1], and the presented ACO strategy has been included in that repository. The flight data used in this work has also been made available online for reproducibility and use by other researchers[2].

5.2   Runtime Environment

All results were gathered using a Beowulf HPC cluster with 32 dual quad-core compute nodes (for a total of 256 processing cores). Each compute node has 64 GBs of 1600 MHz RAM, two mirrored RAID 146 GB 15 K RPM SAS drives, two quad-core E5-2643 Intel processors which operate at 3.3 Ghz, and run the Red Hat Enterprise Linux (RHEL) 6.2 operating system. All 32 nodes within the cluster are linked by a private 56 gigabit (Gb) InfiniBand (IB) FDR 1-to-1 network. The code was compiled and run using MVAPICH2-x [22], to allow highly optimized use of this network infrastructure.

5.3   Data Cleansing

The flight data required some cleaning for use, as it is stored as raw data from the flight data recorders uploaded to the NGAFID server and entered in the database as per second data. When a FDR turns on, some of the sensors are still calibrating or not immediately online, so the first minute of flight data can have missing and erroneous values. These initial recordings were removed from the data the neural networks were trained on. Further, the parameters had wide ranges and different units, e.g., pitch and roll were in degrees, altitude was in meters and airspeed was in knots. These were all normalized to values between 0 and 1 for altitude and airspeed, and -0.5 and 0.5 for pitch and roll.

5.4   Experiments

As backpropagation was shown to not be sufficient to train these recurrent neural networks, particle swarm optimization (PSO) was used to train the neural networks generated by ACO. Previous work has shown both particle swarm and differential evolution as being equally effective in training these networks. PSO used a population of 200, inertia weight of 0.75, and global and local best weights of 1.5 for all runs. PSO was allowed to train the neural networks for 250, 500 and 1000 iterations.

The ACO strategy was used to train networks with 3, 4, and 5 hidden layers (with a similar number of recurrent layers), using 4 and 8 nodes per layer and

---

[1] https://github.com/travisdesell/tao.

[2] http://people.cs.und.edu/~tdesell/ngafid_releases.php.

pheromone degradation rates of 10 %, 5 % and 1 %. The number of ants used was equal to twice the number of nodes per layer (8 for 4 nodes per layer, and 16 for 8 nodes per layer). Each combination of settings was run 5 times for each of altitude, airspeed, pitch and roll, for a total of 1080 runs.

Each run was done allocating 64 processes across 8 nodes, and was allowed to train for 1000 evaluations of generated neural networks. Runs with 250 PSO iterations took around 30 min, 500 PSO iterations took around 1 h, and 1000 PSO iterations took around 2 h.

All runs were done on flight ID 13588 from the NGAFID data release. The neural networks were trained for individual output parameters, as tests have shown that trying to train for multiple output parameters simultaneously performs very poorly, with neither backpropagation or evolutionary strategies being able to find effective weights.

5.5   ACO Parameter Setting Analysis

Figure 3 presents the results of the parameter sweep. In general, there was a strong correlation between increased PSO iterations and the best fitnesses found. Across all runs, 4 nodes per layer performed the best, and apart from altitude, 5 hidden layers performed the best. There did not appear to be a strong trend for the pheromone degradation rate.

5.6   Best Found Neural Networks

Figure 4 displays the best recurrent neural networks evolved by the ACO strategy. For airspeed, pitch and roll, the best networks were the deepest – with 5 hidden and recurrent layers (although all nodes were not used). They also displayed interesting recurrent topologies, significantly different than the standard Jordan and Elman recurrent neural networks found in literature. The best evolved neural network for altitude was also interesting in that it completely ignored roll as an input parameter. The evolved networks also show some slight similarity to sparse autoencoders, with some of the middle layers being constrained to less nodes and connections.

5.7   Comparison to Prior Results

The performance of the best evolved neural networks was compared to the previously best published results for flight ID 13588, which were an Elman network with 2 input lags and 1 hidden layer for airspeed; a Jordan recurrent neural network with 2 input lags and 0 hidden layers for altitude; an Elman network with 1 set put input lags and 1 hidden layer for pitch; and an Elman network with 2 input lags and 1 hidden layer for roll. In addition, results for a random noise estimator (RNE), which uses the previous value as the prediction for the next value, $prediction(t_{i+1}) = t_i$, were given as a baseline comparison, as it represents the best predictive power that can be achieved for random time series data. If the neural networks did not improve on this, then the results would

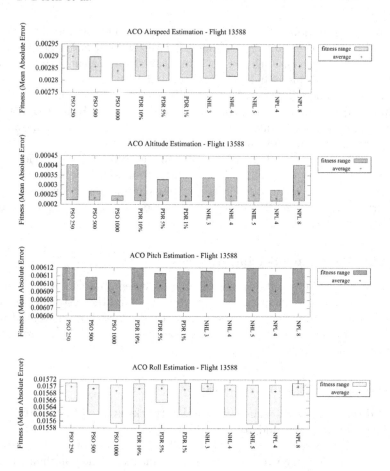

Fig. 3. Minimum, maximum and average fitness (mean average error) given the different ACO input parameters. Fitness values were averaged over each run with the parameter specified in the x-axis. Lower fitness is better. PSO is the number of PSO iterations, PDR is the pheromone degradation rate, NHL is the number of hidden layers, and NPL is the number of nodes per layer.

have been meaningless and potentially indicate that the data is too noisy (given weather and other conditions) for prediction (Fig. 5).

Additionally, the RNE provides a good baseline in that it is easy for neural networks to represent the RNE: all weights can be set to 0, except for a single path from the path from the corresponding input node to the output node having weights of 1. Because of this, it also provides a good test of the correctness of the global optimization techniques, at the very least they should be able to train a network as effective as a RNE; however local optimization techniques (such as backpropagation) may not reach this if the search area is non-convex and the initial starting point does not lead to a good minimum.

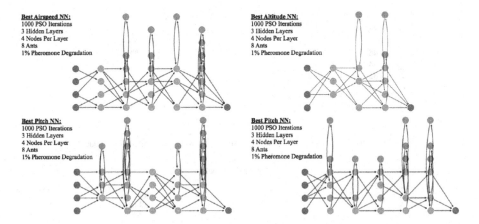

Fig. 4. The best found evolved neural networks across the 1080 runs performed. Input neurons are in blue, recurrent neurons are in pink, hidden neurons are in green, and the output neuron is in purple.

Figure 6 compares the best ACO results to the RNE and the previous best trained neural network for flight ID 13588. Results are the Mean Average Error (MAE) of the prediction to the actual value. As results were normalized over a range of 1, the MAE is also the percentage error. These neural networks and the RNE were also run on four other flights, IDs 15438, 17269, 175755 and 24335 from the NGAFID data release. On average compared to previous best results, the ACO evolved neural networks provided a 63 % improvement over airspeed, a 97 % improvement over altitude and a 120 % improvement over pitch, *without requiring additional input neurons for lag values*. Given the fact that these neural networks also performed strongly on all test flights, these results are quite encouraging.

However, as in previous work, the roll parameter remains quite difficult to predict, and the ACO evolved neural networks actually resulted in a 14.5 % decrease in prediction accuracy, performing worse than the RNE. Given the depth and complexity of the evolved neural networks, there is justifiable concern for over training, which may be the case for this evolved network. Another reason for the poor performance of the ACO evolved neural networks may be due to the limited amount of training for each generated neural network. Previous results had the neural networks be trained for 15,000,000 objective function evaluations, while the best performing ACO evolved neural networks were trained with a maximum of 200,000 objective function evaluations (1000 iterations with population size 200). Given the strong correlation between increased PSO iterations and best fitness found for roll, it is also possible that the neural networks were not trained long enough for the roll parameter. Lastly, it could be that even though the input lag nodes were not required for the other parameters, they may be required for roll, or stand to provide even further prediction improvements. A further study of this stands for future work.

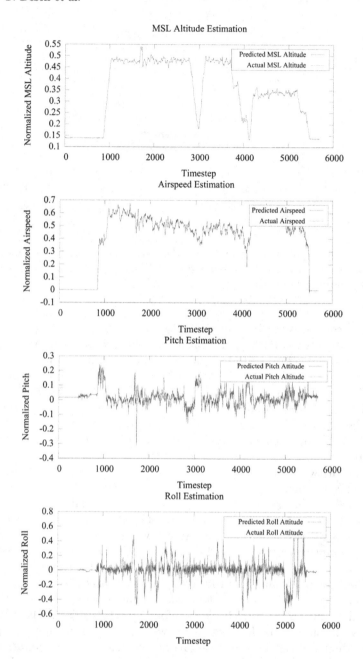

Fig. 5. The best neural networks trained on Flight #13588 were used to predict the parameters of Flight #17269. The actual values are in green and the predictions are in red. Altitude and airspeed were predicted with very high accuracy, however pitch and roll are more challenging. Time steps are in seconds, and parameters are normalized over a range of 1. Predicted and actual airspeed are indistinguishable at the scale of the figure and completely overlap.

Airspeed

Method	13588	15438	17269	175755	24335
$t_{i+1} = t_i$	0.00512158	0.00316859	0.00675531	0.00508229	0.00575537
Prior Best	0.00472131	0.00250284	0.00656991	0.00465581	0.00495454
Best ACO	0.00279963	0.00145748	0.00433578	0.0028908	0.00305361

Altitude

Method	13588	15438	17269	175755	24335
$t_{i+1} = t_i$	0.00138854	0.00107117	0.00200011	0.00137109	0.00192345
Prior Best	0.000367535	0.000305193	0.000895711	0.000399587	0.000485329
Best ACO	0.0002183	0.000160932	0.000353502	0.000224827	0.000249197

Pitch

Method	13588	15438	17269	175755	24335
$t_{i+1} = t_i$	0.0153181	0.010955	0.0148046	0.0161251	0.0173269
Prior Best	0.014918	0.0100763	0.0147712	0.01514	0.0160249
Best ACO	0.00606664	0.00498241	0.00837594	0.005864	0.00733882

Roll

Method	13588	15438	17269	175755	24335
$t_{i+1} = t_i$	0.0158853	0.00604479	0.0204441	0.012877	0.0192648
Prior Best	0.0154541	0.00587058	0.0206536	0.0127999	0.0182611
Best ACO	0.0155934	0.00900393	0.0237235	0.0151416	0.0200261

Fig. 6. Comparison of the best found ACO evolved neural networks to the random noise estimator ($t_{i+1} = t_i$) and the previously published best found results. The mean average error for the neural networks trained on flight ID 13588 is given when they are tested on four other flights.

6 Conclusions and Future Work

This paper presents and analyzes a novel strategy for using ant colony optimization for evolving the structure of recurrent neural networks. The strategy presented is used to evolve deep neural networks with up to 5 hidden and 5 recurrent layers for the challenging task of predicting general aviation flight data, and is shown to provide improvements of 63 % for airspeed, a 97 % for altitude and 120 % for pitch over previously best published results, *while at the same time not requiring additional input neurons for residual values.* Finding good predictions for the roll parameter still remains challenging and an area of future study.

Further, this work opens up interesting opportunites in applying ant colony optimization to neuro evolution. In particular, the authors feel that the approach could be extended to evolve neural networks for computer vision, by allowing ants to also select what type of activation function each neuron has (*e.g.,* ReLU, or max pooling). It may also be possible to utilize this strategy to further improve convolutional layers in neural networks. Additionally, this work only tested neural networks with a recurrent depth of one, where each recurrent node is immediately fed back into the neural network in the next iteration. It may be possible to use this strategy to generate neural networks with deeper memory, where recurrent nodes can potentially feed back into a deeper layer of recurrent nodes, and so on.

Finally, the National General Aviation Flight Database (NGAFID) provides an excellent data source for researching evolutionary algorithms, machine learning and data mining. Further analysis of these flights along with more advanced prediction methods will enable more advanced flight sensors, which could prevent accidents and save lives; which is especially important in the field of general aviation as it is has the highest accident rates within civil aviation [29]. As many of these flights also contain per-second data of various engine parameters, using similar predictive methods it may become possible to detect engine and other hardware failures, aiding in the maintenance process. This work presents a further step towards making general aviation safer through machine learning and evolutionary algorithms.

References

1. Aircraft Owners and Pilots Association (AOPA), January 2014
2. Arenas, M., Collet, P., Eiben, A.E., Jelasity, M., Merelo, J.J., Paechter, B., Preuß, M., Schoenauer, M.: A framework for distributed evolutionary algorithms. In: Guervós, J.J.M., Adamidis, P.A., Beyer, H.-G., Fernández-Villacañas, J.-L., Schwefel, H.-P. (eds.) PPSN 2002. LNCS, vol. 2439, pp. 665–675. Springer, Heidelberg (2002)
3. Ashena, R., Moghadasi, J.: Bottom hole pressure estimation using evolved neural networks by real coded ant colony optimization and genetic algorithm. J. Petrol. Sci. Eng. **77**(3), 375–385 (2011)
4. Bartz-Beielstein, T.: SPOT: an R package for automatic and interactive tuning of optimization algorithms by sequential parameter optimization. arXiv preprint arXiv:1006.4645 (2010)
5. Bilchev, G., Parmee, I.C.: The ant colony metaphor for searching continuous design spaces. In: Fogarty, T.C. (ed.) AISB-WS 1995. LNCS, vol. 993, pp. 25–39. Springer, Heidelberg (1995)
6. Blum, C., Li, X.: Swarm intelligence in optimization. In: Blum, C., Merkle, D. (eds.) Swarm Intelligence, pp. 43–85. Springer, Heidelberg (2008)
7. Blum, C., Socha, K.: Training feed-forward neural networks with ant colony optimization: an application to pattern classification. In: Fifth International Conference on Hybrid Intelligent Systems, 2005, HIS 2005, p. 6. IEEE (2005)
8. Cahon, S., Melab, N., Talbi, E.-G.: Paradiseo: a framework for the reusable design of parallel and distributed metaheuristics. J. Heuristics **10**(3), 357–380 (2004)
9. Chandra, R.: Competitive two-island cooperative coevolution for training elman recurrent networks for time series prediction. In: 2014 International Joint Conference on Neural Networks (IJCNN), pp. 565–572, July 2014
10. Chandra, R., Zhang, M.: Cooperative coevolution of elman recurrent neural networks for chaotic time series prediction. Neurocomputing **86**, 116–123 (2012)
11. Crone, S.F., Hibon, M., Nikolopoulos, K.: Advances in forecasting with neural networks? Empirical evidence from the NN3 competition on time series prediction. Int. J. Forecast. **27**(3), 635–660 (2011)
12. Desell, T.: Asynchronous Global Optimization for Massive Scale Computing. Ph.D. thesis, Rensselaer Polytechnic Institute (2009)

13. Desell, T., Anderson, D., Magdon-Ismail, M., Heidi Newberg, B.S., Varela, C.: An analysis of massively distributed evolutionary algorithms. In: The 2010 IEEE Congress on Evolutionary Computation (IEEE CEC 2010), Barcelona, Spain. July 2010

14. Desell, T., Clachar, S., Higgins, J., Wild, B.: Evolving neural network weights for time-series prediction of general aviation flight data. In: Bartz-Beielstein, T., Branke, J., Filipič, B., Smith, J. (eds.) PPSN 2014. LNCS, vol. 8672, pp. 771–781. Springer, Heidelberg (2014)

15. Desell, T., Szymanski, B., Varela, C.: Asynchronous genetic search for scientific modeling on large-scale heterogeneous environments. In: 17th International Heterogeneity in Computing Workshop, Miami, Florida, April 2008

16. Desell, T., Varela, C., Szymanski, B.: An asynchronous hybrid genetic-simplex search for modeling the Milky Way galaxy using volunteer computing. In: Genetic and Evolutionary Computation Conference (GECCO), Atlanta, Georgia, July 2008

17. Dorigo, M., Birattari, M.: Ant colony optimization. In: Sammut, C., Webb, G.I. (eds.) Encyclopedia of Machine Learning, pp. 36–39. Springer, Boston (2010)

18. Dorigo, M., Gambardella, L.M.: Ant colonies for the travelling salesman problem. BioSystems **43**(2), 73–81 (1997)

19. Dorigo, M., Stützle, T.: Ant colony optimization: overview and recent advances. In: Gendreau, M., Potvin, J.-Y. (eds.) Handbook of Metaheuristics, pp. 227–263. Springer, Boston (2010)

20. Dréo, J., Siarry, P.: A new ant colony algorithm using the heterarchical concept aimed at optimization of multiminima continuous functions. In: Dorigo, M., Di Caro, G.A., Sampels, M. (eds.) Ant Algorithms 2002. LNCS, vol. 2463, pp. 216–221. Springer, Heidelberg (2002)

21. Elias, B.: Securing General Aviation. DIANE Publishing, Darby (2009)

22. Huang, W., Santhanaraman, G., Jin, H.-W., Gao, Q., Panda, D.K.: Design of high performance MVAPICH2: MPI2 over InfiniBand. In: Sixth IEEE International Symposium on Cluster Computing and the Grid, 2006, CCGRID 2006, vol. 1, pp. 43–48. IEEE (2006)

23. Khashei, M., Bijari, M.: A novel hybridization of artificial neural networks and arima models for time series forecasting. Appl. Soft Comput. **11**(2), 2664–2675 (2011)

24. Li, J.-B., Chung, Y.-K.: A novel back-propagation neural network training algorithm designed by an ant colony optimization. In: Transmission and Distribution Conference and Exhibition: Asia and Pacific, 2005 IEEE/PES, pp. 1–5. IEEE (2005)

25. Lukasiewycz, M., Glaß, M., Reimann, F., Teich, J. Opt4j: a modular framework for meta-heuristic optimization. In: Proceedings of the 13th Annual Conference on Genetic and Evolutionary Computation, GECCO 2011, pp. 1723–1730. ACM, New York (2011)

26. MathWorks. Global optimization toolbox. Accessed March 2013

27. Monmarché, N., Venturini, G., Slimane, M.: On how pachycondyla apicalis ants suggest a new search algorithm. Future Gener. Comput. Syst. **16**(8), 937–946 (2000)

28. Mullen, K., Ardia, D., Gil, D., Windover, D., Cline, J.: Deoptim: an R package for global optimization by differential evolution. J. Stat. Softw. **40**(6), 1–26 (2011)

29. National Transportation Safety Board (NTSB) (2012)

30. Ömer Faruk, D.: A hybrid neural network and arima model for water quality time series prediction. Eng. Appl. Artif. Intell. **23**(4), 586–594 (2010)

31. Pandian, A.: Training neural networks with ant colony optimization. Ph.D. thesis, California State University, Sacramento (2013)
32. Shetty, K.I.: Current and historical trends in general aviation in the United States. Ph.D. thesis, Massachusetts Institute of Technology, Cambridge, MA 02139, USA (2012)
33. Sivagaminathan, R.K., Ramakrishnan, S.: A hybrid approach for feature subset selection using neural networks and ant colony optimization. Expert Syst. Appl. **33**(1), 49–60 (2007)
34. Socha, K.: ACO for continuous and mixed-variable optimization. In: Dorigo, M., Birattari, M., Blum, C., Gambardella, L.M., Mondada, F., Stützle, T. (eds.) ANTS 2004. LNCS, vol. 3172, pp. 25–36. Springer, Heidelberg (2004)
35. Socha, K.: Ant Colony Optimisation for Continuous and Mixed-Variable Domains. VDM Publishing, Saarbrücken (2009)
36. Socha, K., Dorigo, M.: Ant colony optimization for continuous domains. Eur. J. Oper. Res. **185**(3), 1155–1173 (2008)
37. Stanley, K., Miikkulainen, R.: Evolving neural networks through augmenting topologies. Evol. Comput. **10**(2), 99–127 (2002)
38. Stanley, K.O., D'Ambrosio, D.B., Gauci, J.: A hypercube-based encoding for evolving large-scale neural networks. Artif. life **15**(2), 185–212 (2009)
39. Szymanski, B.K., Desell, T., Varela, C.A.: The effects of heterogeneity on asynchronous panmictic genetic search. In: Wyrzykowski, R., Dongarra, J., Karczewski, K., Wasniewski, J. (eds.) PPAM 2007. LNCS, vol. 4967, pp. 457–468. Springer, Heidelberg (2008)
40. Unal, M., Onat, M., Bal, A.: Cellular neural network training by ant colony optimization algorithm. In: 2010 IEEE 18th Signal Processing and Communications Applications Conference (SIU), pp. 471–474. IEEE (2010)
41. Ventura, S., Romero, C., Zafra, A., Delgado, J.A., Hervás, C.: JCLEC: a java framework for evolutionary computation. Soft Comput. Fusion Found. Methodol. Appl. **12**(4), 381–392 (2008)
42. Wei, W.W.-S.: Time Series Analysis. Addison-Wesley, Redwood City (1994)
43. Zhang, G.P.: Neural networks for time-series forecasting. In: Armstrong, J.S. (ed.) Handbook of Natural Computing, pp. 461–477. Springer, Boston (2012)

Hyper-heuristic Operator Selection and Acceptance Criteria

Richard J. Marshall[1]([✉]), Mark Johnston[1], and Mengjie Zhang[2]

[1] School of Mathematics, Statistics and Operations Research,
Victoria University of Wellington, Wellington, New Zealand
richardj.marshall@xtra.co.nz, mark.johnston@msor.vuw.ac.nz
[2] School of Engineering and Computer Science,
Victoria University of Wellington, Wellington, New Zealand
mengjie.zhang@ecs.vuw.ac.nz

Abstract. Earlier research has shown that an adaptive hyper-heuristic can be a successful approach to solving combinatorial optimisation problems. By using a pairing of an operator (low-level heuristic) selection vector and a solution acceptance criterion, an adaptive hyper-heuristic can manage development of a "good" solution within an unseen low-level problem domain in a commercially realistic computational time. However not all selection vectors and solution acceptance criteria pairings deliver competitive results when faced with differing problem instance features and computational time limits. We evaluate pairings of six different operator selection vectors and eight solution acceptance criteria, and monitor the performance of the adaptive hyper-heuristic when applying each pairing to a set of Capacitated Vehicle Routing Problem instances of the same size but with different features. The results show that a few pairings of operator selection vector and acceptance criterion perform consistently well, while others require a longer computational time to deliver competitive results. We also investigate some of the features of a problem instance that may influence the performance of the selection vector and acceptance criterion pairings.

1 Introduction

Traditional methods of solving combinatorial optimisation problems use algorithms and heuristics, such as a branch-and-bound algorithm [4] or meta-heuristic search (e.g. tabu search [5]). These methods can achieve good results but often require detailed domain information and can be complex and time consuming to design and execute. A hyper-heuristic is useful where a more general (domain independent) approach is required. The term hyper-heuristic was defined by Cowling et al. [2] as "heuristics to choose heuristics". In this respect, we use a hyper-heuristic to select and execute operators (heuristics) from an unseen set of low-level (domain specific) operators, which in turn incrementally build and/or modify a solution to each problem instance. Understanding what makes a particular hyper-heuristic efficient and effective would enable the trade-off between computational speed and quality of the result to be *managed* when faced with larger problem instances and more complex problem domains.

© Springer International Publishing Switzerland 2015
G. Ochoa and F. Chicano (Eds.): EvoCOP 2015, LNCS 9026, pp. 99–113, 2015.
DOI: 10.1007/978-3-319-16468-7_9

The winning entry of the First Cross-domain Heuristic Search Challenge (CHeSC) [8] in 2011 was an adaptive hyper-heuristic designed by Misir et al. [7]. Marshall et al. [6] illustrate that even a simplified version of the adaptive hyper-heuristic designed by Misir et al. [7] performs well when applied to unseen problems in seven different combinatorial optimisation problem domains. The key components to the adaptive hyper-heuristic are the *operator selection vector*, which determines which operator to apply next, and the *solution acceptance criterion* which determines whether a new solution is retained or discarded. In [6], the simplified adaptive hyper-heuristic used a single operator selection vector and a solution was only accepted if it was better than or equally as good as the solution it may replace. This paper investigates whether providing a wider choice of operator selection vector and solution acceptance criteria can improve the effectiveness of the adaptive hyper-heuristic. We increase the number of possible operator selection vectors to six, and the number of solution acceptance criteria to eight.

The remainder of this paper is organised as follows. A brief background is given in Sect. 2, including an overview of the adaptive hyper-heuristic. Section 3 describes the operator selection vectors and solution acceptance criteria. The experimental design, results and discussion are in Sects. 4 and 5. Finally, Sect. 6 gives our conclusions.

2   Background

This paper uses an adaptive hyper-heuristic (AdaptiveHH) compatible with the HyFlex (**Hy**per-heuristic **Flex**ible) framework [9]. We focus on using a Capacitated Vehicle Routing Problem (CVRP) [12] domain. However, since the hyper-heuristic has no problem domain dependent processes, the hyper-heuristic can readily be applied to other problem domains. Importantly, the hyper-heuristic has no knowledge of the size or features of the problem instance it is working on. This means that the specified computational time limit may be excessive or insufficient to arrive at a "good" solution to the problem instance. The hyper-heuristic must be able to make the best use of the available time and, ideally, terminate processing early when the available time is excessive.

2.1   Adaptive Hyper-heuristic

The AdaptiveHH in this paper is a simplified version of the hyper-heuristic developed by Misir et al. [7]. Conceptually, AdaptiveHH iteratively selects and applies an unseen operator from the problem domain. The resulting solution is then accepted or discarded based on the acceptance criterion specified by AdaptiveHH. AdaptiveHH requires a number of parameters which set the computational time limit, the number of intermediate decision points (phases), and the choice of operator selection vector and acceptance criterion to use (see Fig. 1). Parameters also set the rules about how AdaptiveHH responds if progress towards improving the current solution is stalled.

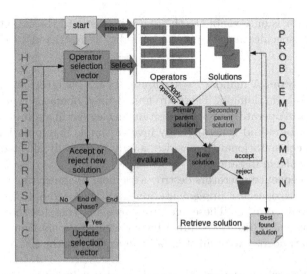

Fig. 1. Overview of how an adaptive hyper-heuristic interacts with a low-level problem domain across the domain barrier.

The two main components of AdaptiveHH are:

1. **The Operator Selection Vector.** This vector is used to select the next operator to apply. The vector is updated at the start of each phase based on the performance of the operator in the preceding phase(s). It gives the probability of each operator being selected and applied to the current solution.
2. **The Solution Acceptance Criteria.** Once an operator modifies a solution to create a new solution, the hyper-heuristic needs to decide whether to accept (retain) or discard the new solution.

2.2   HyFlex Framework

The HyFlex framework [9] was originally developed in 2011 for the First Cross-domain Heuristic Search Challenge (CHeSC) [8]. The framework includes six in-built combinatorial optimisation problem domains. Associated with each in-built problem domain is a set of between 8 and 15 unseen low-level operators (heuristics). Each set contains at least one operator belonging to each of the four defined operator types: *mutation, ruin-recreate, local search* and *crossover*. A crossover operator swaps parts of one solution with another solution in an attempt to create a better solution.

Each operator can use (if appropriate) the two HyFlex parameters α and β, where $(0 \leq \alpha, \beta \leq 1)$. The Intensity of Mutation parameter, α, affects the scale of any mutation or ruin operation, e.g., 0.5 would mean half the current solution would be altered by an operator using this parameter. The Depth of Search parameter, β, defines a range or number of repetitions an operator will undertake to find an improved solution in a single execution of the operator.

Each operator is only visible to the hyper-heuristic to the extent allowed by the HyFlex [9] specifications. Operator visibility is restricted to the following properties:

1. **Operator Type.** A mandatory attribute of each operator contained within a HyFlex problem domain. There are four defined operator types:
   (a) **Mutation** operators add or reposition an element in a solution. Operators of this type would generally only involve simple manipulations requiring a short computational time which is only marginally affected by the size of the problem instance.
   (b) **Ruin-Recreate** operators destroy a segment of an existing solution, chosen by the operator implementation, and then rebuild the segment to form a new solution. These operators are more complex than a mutation operator and typically require a longer computational time. The computational time may vary substantially depending on the size of the problem instance.
   (c) **Local Search** operators define and search a solution neighbourhood for improvements. These operators generally apply a degree of logic to the search so can be expected to have a higher chance of improving a solution, or verify no further local improvements are possible, than the other operator types. However, the computational time may be much longer, and could escalate polynomially (or worse) as the problem instance size increases.
   (d) **Crossover** operators combine elements of two current solutions to form a new solution. The computational time of a crossover operator varies but is often similar to a ruin-recreate operator.
2. **Uses Intensity of Mutation.** An indicator to show whether this operator uses the global Intensity of Mutation, α, parameter.
3. **Uses Depth of Search.** An indicator to show whether this operator uses the global Depth of Search, β, parameter.
4. **Call Record.** The number of times the operator has been executed during a run is calculated and is visible to the hyper-heuristic on demand.
5. **Call Time Record.** The aggregate of the execution time of each operator during a run is recorded and is visible to the hyper-heuristic on demand.

3   The Method

We test the effectiveness of AdaptiveHH by rating each solution generated against the best solution objective value achieved within the computational time limit. We use different pairings of operator selection vector and acceptance criterion. There are 48 possible pairings of operator selection vector (6) and solution acceptance criteria (8).

3.1 Operator Selection Vector Design

AdaptiveHH operates for a specified time limit which is broken down into phases. The operator selection vector is updated at the end of each phase. The choice of selection vector and acceptance criterion is fixed at the beginning of the run and is not altered during the run. The selection vector consists of an array of operators, each with a probability of selection. In the initial selection vector (regardless of type) all operators have an equal probability of selection.

We follow the example of Misir et al. [7] and allow some of the selection vectors described below to exclude operators for one or more phases (i.e. the selection probability is zero). The number of phases an operator is excluded is based on a performance penalty. The first time an operator is excluded the performance penalty is set to one. This means the operator is readmitted to the selection vector at the end of the next phase (i.e. one phase exclusion) with a probability of 0.01 prior to normalisation. If the operator is immediately excluded again during the vector update process at the end of the readmission phase, the performance penalty, and hence the number of exclusion phases, is increased by one. Should the operator be readmitted and survive the vector update process into the succeeding phase, then the performance penalty is reset to one.

The AdaptiveHH reported in [6] used only the Basic Selector. The operator selection vectors are of our own design, but use components of the single selection vector used by Misir et al. [7].

1. **[FS] Fixed Selector:** The initial vector is not altered during the run, so provides a benchmark against which other selection vectors can be measured. All operators have an equal probability of selection regardless of performance.
2. **[BS] Basic Selector:** Updates probabilities by evaluating the success rate of each operator, r_i, since the start of the run:

$$r_i = \frac{\text{number of improvements}_i}{\text{number of calls}_i}$$

   This vector does not exclude operators and sets a minimum probability of selection as 0.001 prior to normalisation.
3. **[P1] Phase Selector (1):** Updates probabilities by evaluating the success rate of the each operator, r_i (as per [BS]), in the most recent phase. During the update process a threshold is set equal to $\frac{1}{3}$ of the success rate of the best performing operator, r_{best}, in that phase. If $r_i \geq \frac{r_{best}}{3}$ it is included in the selection vector for the next phase with a probability of r_i, minimum 0.01, prior to normalisation. Operators where $r_i < \frac{r_{best}}{3}$ are excluded from the vector for the number of phases determined by their performance penalty.
4. **[P2] Phase Selector (2):** Updates probabilities by evaluating the success rate of each operator, r_i (as per [BS]), in the most recent phase. This vector does not exclude operators and sets a minimum probability of selection at 0.001 prior to normalisation. It differs from the Basic Selector in that this selection vector only considers performance during the most recent phase.

5. **[T1] Time Weighted Phase Selector (1):** The time weighted selector uses a time weight, w_i, to penalise slower operators. This is calculated using the average operator execution time, $averageOpTime$, during the preceding phase:

$$w_i = \sqrt{\frac{averageOpTime_i}{averageOpTime_{fastest}}}$$

The time weighted success rate of each operator, r_i, in the most recent phase is evaluated:

$$r_i = \frac{\text{number of improvements}_i}{w_i \times \text{number of calls}_i}$$

During the update process a threshold is set equal to $\frac{1}{3}$ of r_{best}. If $r_i \geq \frac{r_{best}}{3}$ it is included in the selection vector for the next phase with a probability of r_i, minimum 0.01, prior to normalisation. Operators where $r_i < \frac{r_{best}}{3}$ are excluded from the vector for the number of phases determined by their performance penalty.

6. **[T2] Time Weighted Phase Selector (2):** Calculation of the time weight, w_i, and success rates, r_i, are identical to that described for [T1]. For this selector all r_i are ranked highest to lowest, including those excluded from the selection vector ($r_i = 0$). A threshold, T, is set equal to the r_i of the operator ranked $\frac{NumberOfOperators}{2}$ (T may be zero). If $r_i \geq T$, it is included in the selection vector for the next phase with a probability weighting of $\frac{1}{rank}$, prior to normalisation.

3.2   Acceptance Criteria Design

Each application of an operator takes a current solution and modifies it to create a new solution. The new solution is then considered for acceptance into the small population of solutions. If the new solution is not accepted then it is discarded. If the new solution is at least as good as the solution it will replace, then it is automatically accepted into the population of solutions regardless of the acceptance criteria specified by the hyper-heuristic. The following eight acceptance criteria are those proposed by Sabar et al. [10] with minor modifications. As far as possible we have retained the labels and arbitrary parameter values proposed by Sabar et al. [10] and only made changes which are necessary to satisfy the HyFlex framework [9] constraints. In all cases, the new solution is compared to the solution it will replace (if accepted) in the population of solutions.

1. **[IO] Improving or Equal Only:** Only improving (better objective value) or equally good solutions are accepted. All other solutions are discarded.
2. **[AM] Accept Move:** All new solutions are accepted.
3. **[SA] Simulated Annealing:** Non-improving solutions are accepted with a probability $e^{-\delta/t}$, where δ is the change in the objective value between the old and new solutions. The "temperature", t, is $0.5 \times S_{best} \times 0.85^{phase-1}$, where S_{best} is the current best solution objective value [1,11]. The probability of a non-improving solution being accepted decreases as (a) the change in objective value increases and (b) as time progresses.

4. **[MC] Exponential Monte Carlo:** Non-improving solutions are accepted with a probability $e^{-\delta}$, where δ is the change in the objective value between the old and new solutions. The probability of a non-improving solution being accepted decreases as the change in objective value, δ, increases.

5. **[RR] Record to Record Travel:** Non-improving solutions are accepted if the new solution has an objective value less than or equal to $1.03 \times S_{best}$, where S_{best} is the current best solution objective value [3].

6. **[GD] Great Deluge:** Non-improving solutions are accepted if the new solution has an objective value less than or equal to $(1 + 0.85^{phase-1}) \times S_{best}$, where S_{best} is the current best solution objective value [3]. The probability of a non-improving solution being accepted decreases as time progresses.

7. **[NA] Naïve Acceptance:** Non-improving solutions are accepted with 0.5 probability.

8. **[AA] Adaptive Acceptance:** Non-improving solutions are accepted with a probability $1 - \frac{1}{C}$, where $C > 0$ is a counter which increments every 10,000 consecutive operator calls without an improvement in the objective value of the best solution found so far. The counter is reset to 1 each time an improved best solution objective value is found. The probability of a non-improving solution being accepted increases when the search for better solutions reaches a plateau and new best found solutions become harder to find.

4   Experimental Design

The experiments use our own implementation of a CVRP domain [12] compatible with the HyFlex [9] framework. We create 50 random 80-node (79 customers + 1 depot) problem instances requiring a minimum of between 3 and 19 routes. Each problem instance is randomly created using an 80×80 grid. Each instance contains three nodes at fixed locations (see Fig. 2), one of which is the depot, and the other 77 nodes at randomly generated locations. Vehicle capacity is fixed at 1,000 units and each customer's demand is a randomly generated integer with an upper bound ranging from 5 % to 45 % (randomly set for each instance) of the vehicle capacity, with a minimum demand of 1 unit.

We modify the twelve low-level operators proposed by Walker et al. [13] for a CVRP-with-time-windows domain by removing the time window elements from each operator. There are 4 mutation, 2 ruin-recreate, 4 local search and 2 crossover operator types (see Sect. 2.2). The size of the population of solutions is set at six. The hyper-heuristic is only provided with the number of operators of each operator type and has no knowledge of the actual function each operator performs.

We seek to determine:

1. Whether there are particular pairings of operator selection vector and acceptance criterion which consistently perform well or poorly compared to other pairings in arriving at a "good" solution within a short computational time. We examine how each pairing affects the frequency with which each operator type is selected.

2. Whether the location of the depot in relation to the customers influences the consistency and quality of the solutions. To this end we take a problem instance (see Fig. 2) and relocate the depot by swapping the grid coordinates of the depot with one of two customers highlighted. The problem instance is otherwise unchanged. The alternative depot locations are chosen so that the depot is geographically: (a) central, (b) off-centre, and (c) remote.
3. Although we use CVRP instances of the same size, the differing customer demand values mean solutions require a minimum number of routes ranging from 3 to 19. We examine the influence the number of routes has on the performance of the operator selection vector and acceptance criterion pairings. This, and the preceding objective, will determine whether the structure of the problem instance affects performance of the pairings.

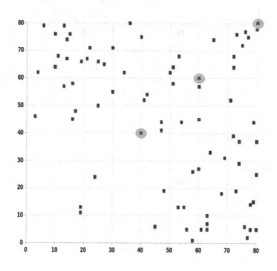

Fig. 2. Randomly generated 80-node problem instance on an 80×80 grid, showing the 3 alternative depot locations (highlighted).

We compare the quality of the results from 30 replications on a set of 50 randomly generated 80 node CVRP instances. We rate individual solutions against the best solution found during the batch of runs (typically 1,440 runs, being 30 replications of 48 pairings) using the following formula (lower ratings are better).

$$\text{rating}_i = \left(\frac{100 \times (\text{solution}_i - \text{solution}_{best})}{\text{solution}_{best}} \right)^2$$

This provides an indication of the relative performance of each pairing compared to its peers. We use the square of the result to increase the apparent difference between results and increase the penalties for poor solutions.

We conduct 25-phase (see Sect. 2.1) experiments using three different depot locations on 50 CVRP instances. We measure pairing performance using computational time limits of 1, 5, 15, 30, 60, 120 and 300 s. The hyper-heuristic we use contains a reinitialisation mechanism if no improving solutions are found for 10,000 consecutive operator calls. It also contains an early termination condition should there still be no improving solutions for two consecutive phases. The purpose of this mechanism is to allow processing to be halted when the hyper-heuristic detects there is a very low likelihood of making further improvements to the best found solution so far.

5   Results and Discussion

Table 1 and Fig. 3 show the results for all pairings from the batches using a 60 s computational time limit for each depot location. Due to space constraints, Tables 2 and 3 only show results from the five best and five worst performing pairings identified in Table 1. Widely differing customer demand values mean the 50 CVRP instances require a minimum of between 3 and 19 routes to service all customers. Table 2 compares the performance of pairings on problem instances where the minimum number of routes is small (3–5 routes), medium (6–13 routes) and large (14–19 routes). Table 3 shows the change in performance over different computational time limits. In Table 3 the performance is measured against the best solution found in any batch for each CVRP instance and depot location. Early terminations only affect the data when allowing a 300 s computational time limit. A negligible number (<0.1 %) of early terminations occurred with a 120 s time limit, and none with the shorter time limits.

Table 1 illustrates the difference in performance when the depot is at different locations. While all pairings provide better results when the depot is located centrally compared to off-centre, the better performing pairs generally

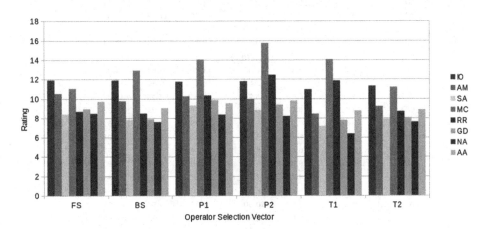

Fig. 3. Comparison of acceptance criteria and selection vector performance ratings (see Sect. 4) during 60 s runs shown in Table 1. Lower ratings are better.

Table 1. Average rating (see Sect. 4) of each selection vector and acceptance criteria pairing over 3 depot locations × 30 replications × 60 s runs on 50 randomly generated CVRP instances (80 nodes, 3 depot locations (see Fig. 2)). Lower ratings are better. Best five performing pairs in bold; five worst in italics.

Acceptance Criteria	Selection Vector	Central Depot	Off-centre Depot	Remote Depot	Average Rating	Std.Dev.
IO	FS	*10.33*	12.35	13.02	11.90	13.62
IO	BS	10.09	12.50	13.04	11.88	13.53
IO	P1	9.96	12.29	12.99	11.75	13.50
IO	P2	10.09	12.29	12.96	11.78	13.38
IO	T1	9.60	11.21	11.99	10.93	12.50
IO	T2	9.74	11.63	12.50	11.29	12.85
AM	FS	9.01	11.89	10.61	10.51	12.42
AM	BS	8.09	11.09	10.02	9.73	11.37
AM	P1	8.19	11.74	10.79	10.24	12.27
AM	P2	7.51	11.57	10.72	9.93	11.97
AM	T1	6.25	10.17	8.86	8.43	10.84
AM	T2	7.35	10.78	9.48	9.20	10.95
SA	FS	6.81	9.25	9.13	8.40	10.31
SA	BS	**5.93**	**8.56**	8.94	7.81	10.36
SA	P1	7.19	10.34	10.36	9.30	11.52
SA	P2	6.74	9.65	10.02	8.80	10.74
SA	T1	**5.59**	**8.06**	**7.83**	**7.16**	9.60
SA	T2	6.10	8.80	8.77	7.89	10.35
MC	FS	9.75	10.96	12.38	11.03	12.54
MC	BS	*11.02*	13.03	*14.70*	*12.92*	14.28
MC	P1	*11.92*	*14.34*	*15.87*	*14.04*	15.28
MC	P2	*13.49*	*16.35*	*17.36*	*15.73*	17.04
MC	T1	*11.77*	*15.01*	*15.37*	*14.05*	15.60
MC	T2	9.99	11.05	12.49	11.18	12.96
RR	FS	6.77	9.46	9.76	8.66	10.88
RR	BS	6.50	9.25	9.63	8.46	10.76
RR	P1	7.73	11.30	11.91	10.31	12.27
RR	P2	9.21	*13.85*	*14.29*	*12.45*	14.21
RR	T1	9.29	*13.05*	13.16	11.83	13.55
RR	T2	6.55	9.30	10.17	8.68	11.01
GD	FS	7.66	9.33	9.81	8.93	11.14
GD	BS	6.47	**7.98**	9.22	7.89	10.29
GD	P1	8.15	10.35	11.00	9.83	11.92
GD	P2	7.57	9.82	10.69	9.36	11.31
GD	T1	6.15	8.74	**8.42**	**7.77**	10.32
GD	T2	6.72	**8.49**	8.88	8.03	10.33
NA	FS	6.71	9.66	8.99	8.45	10.61
NA	BS	**5.96**	8.68	**8.11**	**7.59**	10.01
NA	P1	6.38	9.63	9.04	8.35	10.51
NA	P2	6.33	9.35	8.86	8.18	10.20
NA	T1	**4.58**	**7.60**	**6.95**	**6.37**	8.88
NA	T2	**5.88**	8.64	**8.22**	**7.58**	9.84
AA	FS	8.58	10.60	9.92	9.70	11.58
AA	BS	8.08	9.48	9.51	9.03	10.85
AA	P1	8.28	10.31	9.96	9.52	11.55
AA	P2	8.12	10.74	10.49	9.78	11.91
AA	T1	7.04	9.93	9.26	8.75	11.29
AA	T2	7.62	9.55	9.39	8.85	11.04

show improving performance when the depot is moved even further away from the centre. In contrast, the poorer performing pairings generally show neutral to worsening results the farther the depot is located from the geographic centre. This highlights that the size of the problem instance is not the only factor influencing the performance of the hyper-heuristic. Table 1 also confirms that there is an inter-dependency between the operator selection vector and the acceptance criterion and it is insufficient to separately evaluate each, even though they carry out different functions.

As shown in Table 1, some pairings, such as [SA][T1] and [NA][T1], consistently perform better than other pairings. The pairings using the [MC] and [IO] acceptance criteria generally perform poorly and require a longer computational time to achieve results competitive with the better performing pairings.

A possible cause of this difference is the diversity in the population of solutions. Pairings using the [IO] acceptance criterion, and to a lesser extent [MC], work with a smaller diversity of interim solutions compared to other forms of acceptance criteria. This means that time and effort are not lost on improving low quality solutions that may never become the best solution in the current population of interim solutions. This is a useful trait if the computational time limit is very short, since effort is directed towards improving a better quality solution. On the other hand, accepting only improving or equally good solutions can cause the population of solutions to stagnate and eventually become clones of the best found solution. Once this stage is reached the crossover operators become ineffective and there is a tendency for the process to stall. The hyper-heuristic has a mechanism to reinitialise the population of solutions in the event of stalling, but this is only effective if the selection vector and acceptance criterion pairing can avoid regenerating the same set of solutions.

An increase in the number of routes (see Table 2) as well as the relative second location of the depot are influencing factors as well. However, Tables 1 and 2 also

Table 2. Comparison of the ratings (see Sect. 4) of the five best and five worst performing pairings from Table 1 on CVRP instances requiring a small (15 instances), medium (18 instances) or large (17 instances) minimum number of routes.

Acceptance	Selection	3–5 routes	6–13 routes	14–19 routes
NA	T1	5.84	7.43	5.54
SA	T1	6.30	8.25	6.53
NA	T2	5.30	8.77	7.93
NA	BS	5.30	8.71	8.01
GD	T1	6.41	9.01	7.35
RR	P2	7.94	12.97	15.29
MC	BS	8.24	14.16	15.03
MC	P1	9.56	15.27	16.03
MC	T1	11.84	15.99	12.45
MC	P2	11.55	16.90	17.56

Table 3. Average rating of five best and five worst performing pairings from Table 1 on 50 CVRP instances × 3 depot locations × 30 replications, when allowing a different computational time limit (in seconds). All ratings are measured against the best solution to each instance (and depot location) found during any of the seven batches.

Acceptance	Selection	1	5	15	30	60	120	300
NA	T1	45.27	18.85	12.39	9.25	6.88	5.37	4.07
SA	T1	48.08	20.47	14.14	10.52	7.72	5.88	4.28
NA	T2	888.03	22.97	13.96	10.81	8.19	6.34	4.67
NA	BS	57.72	19.90	13.79	10.89	8.19	6.16	4.49
GD	T1	47.08	22.33	14.97	11.27	10.55	6.40	4.63
RR	P2	46.62	25.11	18.92	15.88	13.29	10.72	8.16
MC	BS	60.19	23.67	18.64	16.06	13.73	11.77	9.69
MC	P1	52.42	29.00	21.62	18.00	14.85	12.35	9.93
MC	T1	54.62	28.95	21.95	18.82	14.88	12.03	9.83
MC	P2	51.74	30.51	23.19	20.01	16.62	13.53	10.96

show that the **relative** performance of operator selection vector and acceptance criterion pairings compared to other pairings is not greatly altered by the number of routes or depot location. A better performing pairing will consistently deliver higher quality solutions than poorer performing pairings regardless of the depot location or the minimum number of routes.

Table 3 shows the performance of each pairing improves with a longer computational time limit, but not all improve at the same rate. The [NA][T2] pairing performs poorly with the 1 s computational time limit but well with longer time limits, indicating a minimum time limit per phase is necessary for some pairings before the operator selection vector update process can be effective. In these experiments the improvement in the performance of the better performing pairings appears to be reaching a plateau with a 300 s time limit. However, the poorer performing pairings show a non-trivial improvement in performance between 120 and 300 s time limits, suggesting a longer computational time may produce further improvements.

Table 4 illustrates how different pairings of operator selection vector and acceptance criterion affect the frequency with which particular operator types are called. The number of calls illustrates how the more aggressive of the two time weighted selection vectors, [T1], biases operator selection towards the faster mutation and crossover operators and away from the slower local search operators. The second time weighted selection vector, [T2], maintains a more balanced selection approach. The Fixed Selector [FS] reflects the 4:2:4:2 balance between the four operator types in the CVRP domain.

The time-weighted selectors favour the faster mutation operators at the expense of the slower search operators. This is a trade-off between speed and quality. Table 4 shows that a large number of operator calls is not critical to the

Table 4. Average number of operator calls, success rate (r_i, as defined in Sect. 3.1) and mix of operator type selection between the six selection vectors (SV) and the Naïve Acceptance [NA], Exponential Monte Carlo [MC] and Improving or Equal Only [IO] acceptance criteria (AC) during experiments using a 60 seconds computational time limit. Best performing pairs (from Table 1) in bold, worst in italics.

AC	SV	Num. calls	Success	Mutation	Ruin-rec.	Search	Crossover
NA	FS	236,900	11.82 %	33.34 %	16.66 %	33.33 %	16.67 %
NA	**BS**	**168,900**	**8.12 %**	**12.65 %**	**15.44 %**	**47.01 %**	**24.89 %**
NA	P1	138,300	5.53 %	5.86 %	10.41 %	54.29 %	29.44 %
NA	P2	231,000	4.71 %	4.40 %	3.57 %	29.40 %	62.63 %
NA	**T1**	**749,400**	**5.78 %**	**2.34 %**	**2.88 %**	**3.53 %**	**91.24 %**
NA	**T2**	**187,500**	**6.82 %**	**17.08 %**	**14.81 %**	**45.22 %**	**22.89 %**
MC	FS	220,800	1.43 %	33.36 %	16.64 %	33.34 %	16.66 %
MC	*BS*	*253,600*	*2.38 %*	*7.11 %*	*5.71 %*	*30.27 %*	*56.91 %*
MC	*P1*	*246,100*	*2.54 %*	*3.41 %*	*2.61 %*	*30.81 %*	*63.17 %*
MC	*P2*	*269,600*	*2.70 %*	*3.30 %*	*1.64 %*	*23.59 %*	*71.47 %*
MC	*T1*	*821,400*	*2.73 %*	*1.68 %*	*0.84 %*	*1.68 %*	*95.80 %*
MC	T2	323,800	1.79 %	51.81 %	11.63 %	22.07 %	14.49 %
IO	FS	229,900	0.16 %	33.32 %	16.66 %	33.35 %	16.67 %
IO	BS	196,100	0.14 %	18.31 %	18.28 %	39.00 %	24.41 %
IO	P1	201,000	0.15 %	25.29 %	18.55 %	35.22 %	20.94 %
IO	P2	191,800	0.16 %	22.32 %	18.20 %	33.23 %	26.25 %
IO	T1	512,100	0.18 %	34.95 %	8.69 %	8.55 %	47.81 %
IO	T2	373,200	0.16 %	56.22 %	12.51 %	18.85 %	12.41 %

quality of the solution. This table also illustrates the lower number of calls made to crossover type operators when the [IO] acceptance criteria is used, reflecting the reduced effectiveness of these operators in this situation. In contrast, the time weighted selector [T1] almost exclusively uses the crossover operator with both the Naïve Acceptance [NA] and Exponential Monte Carlo [MC] acceptance criteria. With [NA], the resulting solutions are among the best, while with [MC] they are among the worst. This can be explained by the difference in the diversity of the population of solutions, as reflected in the relative operator call success rates. However other factors such as parameter values and the number of early terminations (42 % during 300 s time limit) due to best found solutions no longer improving, may also influence the difference in overall solution quality.

6   Conclusions

When comparing the results in Tables 1, 2 and 3 we deduce the following about the operator selection vector and acceptance criteria pairings.

1. Generally perform well:
   (a) Time Weighted Phase Selectors [T1] and [T2] with the Naïve Acceptance [NA] criterion.
   (b) Time Weighted Phase Selector (1) [T1] with Simulated Annealing [SA] acceptance criterion.
2. Generally perform poorly: Any operator selection vector with:
   (a) The Exponential Monte Carlo acceptance criterion [MC].
   (b) The Improving or Equal Only acceptance criterion [IO].

Correctly setting the early termination criteria means the hyper-heuristic can determine the correct computational time even though it has no knowledge of the problem instance size or features. We propose to undertake further work on this feature.

In future research we shall examine whether the relative performance of the operator selection vector and acceptance criterion pairings is consistent across problem instances of differing sizes. We shall also evaluate the merits of enabling the adaptive hyper-heuristic to change the pairing of operator selection vector and acceptance criteria during an interim phase update.

References

1. Bai, R., Blazewicz, J., Burke, E., Kendall, G., McCollum, B.: A simulated annealing hyper-heuristic methodology for flexible decision support. 4OR Q. J. Oper. Res. **10**(1), 43–66 (2012)
2. Cowling, P., Kendall, G., Soubeiga, E.: A hyperheuristic approach to scheduling a sales summit. In: Burke, E., Erben, W. (eds.) Practice and Theory of Automated Timetabling III. LNCS, vol. 2079. Springer, Heidelberg (2000)
3. Dueck, G.: New optimization heuristics: the great deluge algorithm and the record-to-record travel. J. Comput. Phys. **104**(1), 86–92 (1993)
4. Fisher, M.: Optimal solution of vehicle routing problems using minimum k-trees. Oper. Res. **42**(4), 626–642 (1994)
5. Glover, F.: Tabu search: Part I. ORSA J. Comput. **1**(3), 190–206 (1989)
6. Marshall, R.J., Johnston, M., Zhang, M.: A comparison between two evolutionary hyper-heuristics for combinatorial optimisation. In: Dick, G., Browne, W.N., Whigham, P., Zhang, M., Bui, L.T., Ishibuchi, H., Jin, Y., Li, X., Shi, Y., Singh, P., Tan, K.C., Tang, K. (eds.) SEAL 2014. LNCS, vol. 8886, pp. 618–630. Springer, Heidelberg (2014)
7. Misir, M., Verbeeck, K., De Causmaecker, P.: A new hyper-heuristic as a general problem solver: an implementation in HyFlex. J. Sched. **16**, 291–311 (2013)
8. Ochoa, G., Hyde, M.: Cross-domain Heuristic Search Challenge (2011). http://www.asap.cs.nott.ac.uk/external/chesc2011/
9. Ochoa, G., Hyde, M., Curtois, T., Vazquez-Rodriguez, J.A., Walker, J., Gendreau, M., Kendall, G., McCollum, B., Parkes, A.J., Petrovic, S., Burke, E.K.: HyFlex: a benchmark framework for cross-domain heuristic search. In: Hao, J.-K., Middendorf, M. (eds.) EvoCOP 2012. LNCS, vol. 7245, pp. 136–147. Springer, Heidelberg (2012)
10. Sabar, N., Ayob, M., Kendall, G., Qu, R.: Grammatical evolution hyper-heuristic for combinatorial optimization problems. IEEE Trans. Evol. Comput. **17**(6), 840–861 (2013)

11. Soubeiga, E.: Development and application of hyperheuristics to personnel scheduling. Ph.D thesis, University of Nottingham (2003)
12. Toth, P., Vigo, D.: The Vehicle Routing Problem. SIAM. LNCS. Springer, Philadelphia (2002)
13. Walker, J.D., Ochoa, G., Gendreau, M., Burke, E.K.: Vehicle routing and adaptive iterated local search within the *HyFlex* hyper-heuristic framework. In: Hamadi, Y., Schoenauer, M. (eds.) LION 2012. LNCS, vol. 7219, pp. 265–276. Springer, Heidelberg (2012)

Improving the Performance of the Germinal Center Artificial Immune System Using ϵ-Dominance: A Multi-objective Knapsack Problem Case Study

Ayush Joshi[✉], Jonathan E. Rowe, and Christine Zarges

School of Computer Science, University of Birmingham,
Edgbaston, Birmingham, UK
{axj006,j.e.rowe,c.zarges}@cs.bham.ac.uk

Abstract. The Germinal center artificial immune system (GC-AIS) is a novel immune algorithm inspired by recent research in immunology, which requires very few parameters to be set by hand. The population of solutions in GC-AIS is dynamic in nature and has no restrictions on its size which can cause problems of population explosion, where the population keeps growing very rapidly, leading to wasteful fitness evaluations. In this paper we try to address this problem in the GC-AIS by incorporating ϵ-dominance, which is a well known mechanism in multi-objective optimization to regulate population size. The improved variant of GC-AIS is compared with a well known multi-objective evolutionary algorithm NSGA-II on the multi-objective knapsack problem. We show that our improved GC-AIS performs better than NSGA-II on the instances of the knapsack problem taken from [23] inheriting the same benefits of having to set fewer parameters manually.

Keywords: Artificial immune systems · GC-AIS · NSGA-II · Knapsack problem

1 Introduction

Multi-objective optimisation is the task of finding optimal solutions to a problem which has several objectives, that often compete with each other. Hence, there exists a set of optimal solutions to the problem called the pareto optimal set. Real life problems often involve competing objectives and the complexity of these problems can cause problems when applying exact methods [9]. Evolutionary algorithms have been employed to solve multi-objective optimization problems since the mid-1980s [15] with good success.

Artificial immune systems (AIS) are meta-heuristics which have been developed taking inspiration from different models of the immune system of vertebrates. The immune system is an especially interesting biological system as it possesses several desirable properties combined together. Due to features

© Springer International Publishing Switzerland 2015
G. Ochoa and F. Chicano (Eds.): EvoCOP 2015, LNCS 9026, pp. 114–125, 2015.
DOI: 10.1007/978-3-319-16468-7_10

like diversity, robustness, and memory, AIS have been applied to a large number of applications such as machine learning, security, robotics, optimization Castro and Timmis [2] provide a detailed survey of applications. Compared with other optimization algorithms AIS are relatively new, and a survey of AIS that have been applied to solve multi-objective optimization problems is provided by Freschi et al. [5].

Natural processes perform several complicated tasks with efficiency and tend to be robust. Current approaches used to solve real world problems are facing problems with robustness and scalability [6,11], as real world problems become more and more complex. Therefore it is a growing trend that understanding and using more detailed ideas from natural processes can help us design better performing systems. Towards this some work has been done by Greensmith et al. [7] but this has mostly been limited to intrusion detection and classification. Sim et al. [17] have proposed a hyper-heuristic called NELLI, which learns from changing problem landscapes, and has been shown to perform better than single human-designed heuristics.

GC-AIS is a novel AIS introduced by Joshi et al. [10], which is inspired by recent research on the Germinal center reaction [19]. It has interesting properties like dynamic population size and it requires very few parameters to be manually specified. In each generation of the GC-AIS every solution creates a mutated clone which can potentially cause population explosion. This characteristic can sometimes cause problems as many fitness evaluations are wastefully expended without observing significant improvements in solutions.

In this paper we replace the dominance comparison in the GC-AIS with ε-dominance as a method to control population size. This modified GC-AIS which we call ε-GC-AIS is compared with the non-dominated sorting genetic algorithm-II (NSGA-II) [3] on instances of the multi-objective d-dimensional knapsack problem (MOd-KP) taken from [23]. Three measures used in multi-objective optimization, namely hypervolume [21], generational distance [18] and generalized spread [20] are used as metrics for comparison. It is shown that with a suitable choice of ε value, ε-GC-AIS performs better than NSGA-II and has the inherent benefit of requiring less parameters to be set manually.

The outline of the paper is as follows: In Sect. 2 multi-objective optimization is introduced along with a detailed description of the multi-objective knapsack problem. Section 3 gives the description of the GC-AIS model along with a description of the ε-GC-AIS algorithm. In Sect. 4 the experimental setup is explained and along with the obtained results in Sect. 5. The paper is concluded in Sect. 6 with a discussion on the observed results and conclusions thereafter.

2   Preliminaries

2.1   Multi-objective Optimization

Multi-objective optimization can be described as finding solutions to problems which have more than one objective, that are often in competition with each other. If we denote the fitness of the different objectives as f_i $(1 \leq i \leq n)$, then a solution x^1 dominates another solution x^2 if $\forall i\ f_i(x^1) \geq f_i(x^2)$ and there exists at least one i such that $f_i(x^1) > f_i(x^2)$, assuming maximization without loss of generality. At any time during its run, a multi-objective optimization algorithms may maintain a set of non-dominating solutions to the problem. If for some solution x there exists no other solution y that dominates x, then x is a pareto optimal solution to the problem. A pareto optimal set is a set containing all pareto optimal solutions.

2.2   The Multi-objective D-Dimensional Knapsack Problem

The knapsack problem is a widely studied NP-hard combinatorial optimization problem [23] with real world applications like capital budgeting and resource allocation [16]. The single objective knapsack problem consists of a set of items which have associated weights and profits, and a knapsack which has a fixed capacity. The goal is to find the set of items which can be packed in the knapsack giving the maximum profit without exceeding the capacity of the knapsack. By introducing multiple knapsacks this single objective problem can be extended to a multi-objective version. Each knapsack has its own capacity and the items have different profits and weights associated with each knapsack. A practical example of this problem is packet scheduling for wireless networks with relay nodes [1].

The solutions to the multi-objective d-dimensional knapsack problem (MOd-KP) can be encoded as bit strings of length m where m is the total number of items available. A 1 indicates the presence of an item while 0 indicates the absence of the item. It should be noted that if a bit is 1 then an item is considered to be present in all the knapsacks and vice-versa. A more formal definition of the multi-objective knapsack problem can be stated as:

Definition 1. *Let m denote the number of items and n the number of knapsacks. Let $p_{i,j}$ and $w_{i,j}$ the profits and weights of item i with respect to knapsack j, respectively, and c_j the capacity of knapsack j. Let a solution be represented as $x = (x_1, x_2, \cdots, x_m)$ with $x_i \in \{0,1\}$. The objective is to maximize*

$$f(x) = (f_1(x), f_2(x), \cdots, f_n(x))\ where\ f_i(x) = \Sigma_{j=1}^m p_{i,j} \cdot x_i$$

subject to $\Sigma_{j=1}^n w_{i,j} \cdot x_j < c_i; \forall i \in \{1, 2, 3, \ldots n\}$.

The MOd-KP is a constrained multi-objective problem where the constraint is the capacity of the knapsacks. Not all possible bit string combinations represent valid solutions and some repair mechanism must be applied in order to transform invalid solutions to valid ones. In [23], the authors used the maximum profit by weight ratio (greedy repair) method to repair invalid solutions. For an item i, the maximum profit by weight ratio is given by:

$$q_i = \max(p_{i,j}/w_{i,j}), \quad \forall j \in (1, 2, \ldots, n). \tag{1}$$

An item with the lowest maximum profit by weight ratio is removed first, and they are removed iteratively in an increasing order of the ratio, until a feasible solution is obtained.

Another repair heuristic was introduced in [9] where the weighted profit by weight ratio (weighted scalar repair), is used to find the order of removal of items. In this approach the items are sorted based on:

$$q_i = \left(\sum_{j=1}^{n} \lambda_i p_{i,j}\right) / \left(\sum_{j=1}^{n} w_{i,j}\right), \quad \forall j \in (1,2,\ldots,n), \tag{2}$$

where the λ_i are the scalar coefficients of the linear utility function used for scalarizing the multi-objective fitness vector in the Multi-objective genetic local search (MOGLS) [9] algorithm. These coefficients are generated randomly at each generation of MOGLS and are used for selection as well as repair. The generation procedure for normalized weight vectors is given in Algorithm 1.

Algorithm 1. Algorithm for generation of normalized weight vectors. rand returns a random number between 0 and 1. [9]

$\lambda_1 = 1 - \sqrt[n-1]{rand()}$

. . .

$\lambda_j = (1 - \sum_{i=1}^{j-1} \lambda_i) \cdot (1 - \sqrt[j-1]{rand()})$

. . .

$\lambda_n = 1 - \sum_{i=1}^{n-1} \lambda_i$

---

3   The ϵ-GC-AIS Algorithm

The ϵ-GC-AIS is an improved variant of GC-AIS originally proposed by Joshi et al. [10], which is a new AIS designed using knowledge from cutting edge research in immunology, specifically the understanding of the germinal center (GC) reaction [19]. Germinal centres are regions in the immune system where B cells a type of immune cell that generates antibodies (*Abs*) to fight an infection, are presented with the invading pathogens [14].

The key highlights of the GC reaction are as follows: At the start of the invasion the number of GCs grows and they try to find the best *Ab* for the pathogen by continuously mutating and selecting the B cells which can bind with the pathogen. There is periodic communication between GCs by transmitting *Abs*. By proliferation, mutation and selection of B cells this reaction is able to produce *Abs* which can eradicate the pathogen. Towards this stage the number of GCs starts declining. The new theory proposed in [19] deals with the selection aspect of this reaction, where there is a direct competition between *Abs* and mutating B cells, and the cells which are unable to compete die by apoptosis (cell death). Entire GCs may disappear if the B cells within them cannot compete with *Abs* from neighbouring GCs.

Based on Algorithm 2 a brief description of ϵ-GC-AIS is as follows. The ϵ-GC-AIS starts with one GC which contains one individual that represents a B cell.

Algorithm 2. The ϵ-GC-AIS

Let G^t denote the population of GCs at generation t and g_i^t the i-th GC in G^t.
Create GC pool $G^0 = \{g_1^0\}$ and initialise g_1^0. Let $t := 0$.
loop
    for each GC g_i^t in pool G^t in parallel **do**
        Generate random weight vectors λ_i 1.
        Create offspring y_i of individual g_i^t by standard bit mutation.
        Repair invalid offsprings using the weighted scalar repair approach.
    end for
    Add all y_i to G^t, remove all ϵ-dominated solutions from G^t and let $G^{t+1} = G^t$.
    Let $t = t + 1$.
end loop

Offspring are created by standard bit mutation of B cells in GCs. Standard bit mutation means that each bit can be flipped with the probability $1/n$. In every generation there is a migration of fitness values of the offspring between GCs, which corresponds to the migration of Ab between GCs. After migration, ϵ-dominated solutions are deleted which correspond to the eradication of GCs. The surviving offspring correspond to new GCs. This leads to a model where the number of GCs is dynamic in nature. The difference in this modified GC-AIS from the original model is the incorporation of ϵ-dominance which replaces the previous standard dominance relation. The ϵ-GC-AIS always maintains a set of ϵ-non-dominated solutions in every generation and the population size of GCs is dynamic in nature.

ϵ-dominance [12] is a generalization of the dominance relation, which can be conceptually visualized for the two dimensional case as follows: the objective space is divided into a grid of rectangles of dimensions ϵ and solutions are mapped onto this grid. Assuming maximization, if two solutions are in different boxes then they are compared using the standard dominance relation, but if they are in the same box then they are compared based on their euclidean distance from the origin, and the solution with the larger distance is kept. Figure 1 gives a diagram of the ϵ dominated front, where the red circles represent the dominated points and the green circles represent the point on the ϵ dominated front.

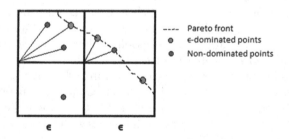

Fig. 1. Diagram of ϵ dominated front along with dominated points

4   Experimental Setup

In this section we describe the experimental setup used in this study: ϵ-GC-AIS and the NSGA-II are compared by testing them on some benchmark instances of the multi-objective knapsack problem. A total of 12 instances are provided in [23] and are grouped into 3 classes based on the number of knapsacks. Each class consists of 4 instances based on the number of total items. The different knapsack numbers are 2, 3 and 4 and the number of items are 100, 250, 500 and 750. 30 independent runs each with one million fitness evaluations were performed for each algorithm and statistical results were recorded. Any-time data and end-of-run performance metrics are provided.

NSGA-II [3] is a widely used elitist multi-objective evolutionary algorithm which uses non-dominated sorting to rank solutions according to levels of dominated fronts, as well as crowding distance measure to maintain diversity in the solution set. In each generation a fixed population size N is maintained which is initialized by the user, individuals are selected by tournament selection using the ranks and crowding distance and are then subject to crossover and mutation to create N offspring. These offspring are combined with the parent population and the best N solutions are carried to the next generation based on the crowding distance measure and non-domination.

In the work by Zitzler et al. [22], authors state that NSGA-II and SPEA2 have similar behaviour on different problems considered in their work which includes MOd-KP, where they have used the greedy approach for repair. Ishibuchi and Kaige [8] show that the weighted scalar repair approach improves diversity of solutions and also increases convergence speed in many cases. It is also shown that performance of multi-objective algorithms strongly depends on the choice of the repair procedure used which is confirmed by work done in [16]. The implementation of NSGA-II in this paper closely resembles that of the work in [8], and the algorithm is kept as close to its pure form as possible with the inclusion of the weighted scalar repair. One random weight vector is generated for every solution which needs repair. In this work the implementation of NSGA-II has been adapted from the MOEA framework[1].

The ϵ-GC-AIS algorithm requires only two parameters to be set by hand, the mutation rate and a value for ϵ, on the other hand NSGA-II requires three parameters to be set, namely population size, the mutation rate and probability of crossover. For our experiments these settings are similar to the ones used by [16] as our implementation of NSGA-II is closest to their ϵ-NSGA-II where all the parameters values have been kept as close to the original work in [22,23]. The mutation rate for NSGA-II has been set to $5/n$ according to the guidelines in [13]. The mutation rate is the probability with which every bit is flipped per individual, while mutation probability refers to the probability of each individual to be mutated. The parameter values can be seen in Table 1. Since population size for NSGA-II is usually set according to the problem instance, the values again have been selected from [16] and are depicted in Table 2.

---

[1] www.moeaframework.org.

Table 1. Algorithm settings for GC-AIS and NSGA-II

Algorithm parameters	NSGA-II	ϵ-GC-AIS
Initial population	Refer Table-2	1
Selection	Tournament	All reproduce
Mutation type	Standard bit mutation	Standard bit mutation
Mutation probability	1	1
Mutation rate	5/n	1/n
Crossover type	One point crossover	-
Crossover probability	0.8	-
Epsilon value	-	20/50

Table 2. Population size used in NSGA-II for different instances of MOd-KP

Items	Knapsacks		
	2	3	4
100	150	200	250
250	150	200	250
500	200	250	300
750	250	300	400

For the purpose of comparison three popular measures used in evolutionary multi-objective optimization, the hypervolume metric [22], generational distance [18] and the generalized spread metric [20], have been used. The spread metric is a measure of how the solutions are distributed in the non-dominated front. This measure was proposed for problems with two dimensions in [3] but was later generalized for any number of dimensions in the work by [20]. The lower the value for the metric the better the spread. The hypervolume metric is a measure of the volume dominated by the achieved non-dominated set with respect to a nadir point, and higher values of this metric are considered better. The generational distance is a method to estimate how far the solutions in the non-dominated set are from the true pareto set, and lower values for this metric are considered to be better.

The implementations of these metrics have been adapted from the Jmetal framework [4] where, for generational distance and spread, the non-dominated front as well as known best reference fronts are required. The authors of [23] have provided the pareto fronts for only 4 instances, namely the 2 knapsack instance with items 100, 250 and 500 and 3 knapsack with items 100. Therefore, the technique employed in [16] has been used to generate best-known non-dominated fronts for the remaining instances. According to this method, ϵ-GC-AIS and NSGA-II were run for 10 independent runs each for 1 million fitness evaluations. The final non-dominated fronts from each run were combined and all dominated solutions were removed. The resulting fronts were used as reference fronts for the generation of performance measures.

5   Results and Discussion

The first set of experiments is performed to find a suitable value for ϵ. Five different values of ϵ were tried namely 2, 5, 10, 15, and 20 along with standard GC-AIS with dominance without ϵ and these were compared with NSGA-II. Run-time plots of population, hypervolume, generational distance and spread are provided in Fig. 2. 30 independent runs were performed for each setting and plots of the mean of population, hypervolume, generational distance and spread metric for each generation averaged over 30 runs were plotted. A representative plot for the two dimensional problems is provided here. Including curves for all ϵ values was not possible in Fig. 2, and selected values of 10, 20, along with NSGA-II and standard GC-AIS ($\epsilon = 0$) are shown.

It can be seen from Fig. 2 that early in the run the GC-AIS obtains better hypervolume than NSGA-II using all values of ϵ, and NSGA-II catches up around the middle. The plots for spread measure show that using different ϵ values can cause variations in the spread. It can be seen from the plots of hypervolume and generational distance that ϵ-GC-AIS achieves better values in the early phase of the runs, while NSGA-II takes some time to catch up. Also, the plots for the

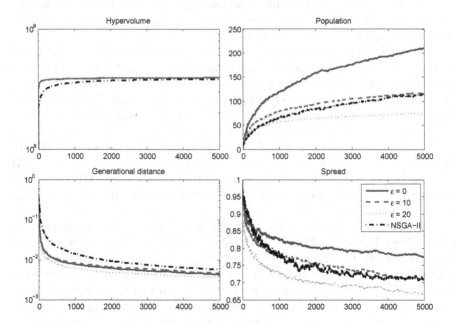

Fig. 2. Any-time plots for ϵ-GC-AIS and NSGA-II. X axes show the generations, Y axes show the measure of the respective metrics. Instance size 500 items, 2 knapsacks, Hypervolume and Generational distance are plotted on log scale.

Table 3. Hypervolume measure values for NSGA-II and ϵ-GC-AIS

Sacks	Items	ϵ-GC-AIS average	std.dev	NSGA-II average	std.dev	Wilcoxon test
2	100	1.68e+7	6.26e+4	1.68e+7	5.37e+4	0.137
	250	**9.59e+7**	3.57e+5	9.51e+7	3.85e+5	5.92e-8
	500	**3.94e+8**	1.08e+6	3.83e+8	1.72e+6	3.01e-11
	750	**8.51e+8**	3.12e+6	8.19e+8	3.24e+6	3.01e-11
3	100	6.11e+10	4.68e+8	6.08e+10	3.79e+8	0.079
	250	**8.46e+11**	6.29e+9	8.29e+11	5.28e+9	1.32e-10
	500	**6.56e+12**	3.87e+10	6.42e+12	3.89e+10	3.33e-11
	750	**2.19e+13**	1.19e+11	2.15e+13	1.80e+11	3.01e-11
4	100	1.51e+14	2.77e+12	**1.63e+14**	1.17e+12	3.01e-11
	250	6.33e+15	6.78e+13	**6.44e+15**	6.15e+13	1.15e-7
	500	9.72e+16	9.81e+14	9.72e+16	1.05e+15	0.371
	750	4.70e+17	3.91e+15	**4.82e+17**	4.80e+15	4.61e-10

population size show that ϵ value has a big impact on the population size. Since we are interested in preventing population explosion, which was observed to be especially common in the large dimensional instances, an ϵ value of 20 is selected for 2 and 3 dimensions as it produces the lowest population and the best spread. For the 4 dimensional problems an ϵ value of 50 is used, since the population size becomes quite large for these instances with $\epsilon = 20$.

The end-of-run performance is calculated next, where the metrics are recorded after each run. Averages of metrics have been recorded along with standard deviation. Wilcoxon rank-sum test was performed as a statistical measure to ascertain the level of significant difference between the algorithms. These values for the hypervolume metric are shown in Table 3, for the spread metric are shown in Table 4 and for the generational distance in Table 5.

It can be seen from Table 3 that in six out of the twelve instances ϵ-GC-AIS achieves greater hypervolume than NSGA-II. While on three instances NSGA-II performs better than ϵ-GC-AIS. In all but three instances the Wilcoxon rank sum test was able to confirm significant difference between the two algorithms at a significance level of 0.05.

Values from the spread metric from Table 4 show that the non-dominated fronts achieved by ϵ-GC-AIS are better distributed than the ones achieved by NSGA-II. Except for one instance, the wilcoxon rank sum test confirmed significant difference between the two algorithms. Similar results can be seen from Table 5 where for each of the instances it can be seen that the generational distance measure for the NSGA-II is higher than ϵ-GC-AIS which means that the non-dominated fronts achieved by ϵ-GC-AIS are closer to the reference fronts than those of NSGA-II. In this case the wilcoxon rank sum test showed significant difference between the two algorithms for all the instances.

Table 4. Spread measure values for NSGA-II and ϵ-GC-AIS

		ϵ-GC-AIS		NSGA-II		
Sacks	Items	Average	Std.dev	Average	Std.dev	Wilcoxon test
2	100	**0.507**	0.037	0.568	0.056	5.09e-6
	250	0.671	0.024	0.668	0.050	0.923
	500	**0.666**	0.027	0.710	0.040	5.26e-5
	750	**0.493**	0.044	0.645	0.031	3.01e-11
3	100	**0.350**	0.020	0.388	0.027	5.59e-7
	250	**0.335**	0.011	0.439	0.026	3.01e-11
	500	**0.363**	0.010	0.455	0.025	3.33e-11
	750	**0.394**	0.010	0.471	0.025	3.68e-11
4	100	**0.298**	0.018	0.361	0.020	5.49e-11
	250	**0.317**	0.011	0.406	0.025	3.33e-11
	500	**0.304**	0.007	0.440	0.020	3.01e-11
	750	**0.323**	0.006	0.454	0.021	3.02e-11

Table 5. Generational distance measure for ϵ-GC-AIS and NSGA-II

		ϵ-GC-AIS		NSGA-II		
Sack	Items	Average	Std.dev	Average	Std.dev	Wilcoxon test
2	100	**0.0010**	3.93e-4	0.0038	6.37e-4	3.01e-11
	250	**0.0015**	1.75e-4	0.0071	6.86e-4	3.01e-11
	500	**0.0029**	2.56e-4	0.0137	8.62e-4	3.01e-11
	750	**0.0013**	2.70e-4	0.0166	0.0010	3.01e-11
3	100	**0.0015**	1.51e-4	0.0079	5.82e-4	3.01e-11
	250	**9.67e-4**	1.51e-4	0.0146	0.0013	3.01e-11
	500	**8.43e-4**	1.52e-4	0.0180	0.0011	3.01e-11
	750	**8.41e-4**	1.27e-4	0.0117	8.14e-4	3.01e-11
4	100	**0.0028**	2.73e-4	0.0142	8.89e-4	3.00e-11
	250	**0.0011**	1.46e-4	0.0199	0.0012	3.01e-11
	500	**7.64e-4**	8.35e-5	0.0182	0.0011	3.01e-11
	750	**7.92e-4**	1.57e-4	0.0144	7.18e-4	3.01e-11

6   Conclusion

It can be seen from Fig. 2 that setting the right ϵ value is important for maintaining the population size in the ϵ-GC-AIS. Along with regulating the population size, it provides the added advantage of maintaining diversity between the solutions. Parameter setting is a crucial factor when employing any meta-heuristic to solve a problem. It can be seen from Table 1 that ϵ-GC-AIS requires the ϵ

parameter as well as mutation rate to be set while NSGA-II requires population size, as well as mutation rate and crossover probability.

We have shown that ϵ-GC-AIS performs better than NSGA-II on the MOd-KP instances provided by [23]. The value of ϵ has a big impact on the population size of ϵ-GC-AIS, which can be seen from Fig. 2. In the higher dimensional cases the number of non-dominated solutions increases very rapidly and possibly larger ϵ values could further improve hypervolume in those instances in Table 3. ϵ is a new parameter which was not present in the original description of GC-AIS. We would like to remove the task of setting ϵ manually by incorporating some form of dynamic ϵ resizing, where this value will dynamically adjust according to population size and the number of fitness evaluations expended so far, and also do this for each dimension. This is a direction for future work.

As future work we would also like to incorporate the correlated instances of MOd-KP as suggested in [16] and test the performance on the many objective knapsack problem. Though NSGA-II is a popular multi-objective evolutionary algorithm and with incorporation of the weighted repair approach can be considered as one of the state of the art MOEAs, we would like to include other heuristics and MOEAs for future studies for MOd-KP.

References

1. Cohen, R., Grebla, G.: Multi-dimensional OFDMA scheduling in a wireless network with relay nodes. In: INFOCOM, pp. 2427–2435. IEEE (2014)
2. De Castro, L.N., Timmis, J.: Artificial Immune Systems: A New Computational Intelligence Approach. Springer, Heidelberg (2002)
3. Deb, K., Agrawal, S., Pratap, A., Meyarivan, T.: A fast elitist non-dominated sorting genetic algorithm for multi-objective optimisation: NSGA-II. In: PPSN VI, pp. 849–858. Springer (2000)
4. Durillo, J., Nebro, A., Alba, E.: The jMetal framework for multi-objective optimization: design and architecture. In: CEC, pp. 4138–4325. Springer, July 2010
5. Freschi, F., Coello, C.A.C., Repetto, M.: Multiobjective optimization and artificial immune systems: a review. Handb. Res. Artif. Immune Syst. Nat. Comput. Applying Complex Adapt. Technol. **4**, 1–21 (2009)
6. Greensmith, J.: The dendritic cell algorithm. Ph.D. thesis, University of Nottingham (2007). http://www.cs.nott.ac.uk/~qg/thesis.pdf
7. Greensmith, J., Aickelin, U., Twycross, J.: Articulation and clarification of the dendritic cell algorithm. In: Bersini, H., Carneiro, J. (eds.) ICARIS 2006. LNCS, vol. 4163, pp. 404–417. Springer, Heidelberg (2006)
8. Ishibuchi, H., Kaige, S.: Effects of repair procedures on the performance of EMO algorithms for multiobjective 0/1 knapsack problems. In: CEC, vol. 4, pp. 2254–2261. IEEE (2003)
9. Jaszkiewicz, A.: On the performance of multiple-objective genetic local search on the 0/1 knapsack problem - a comparative experiment. IEEE Trans. Evol. Comput. **6**(4), 402–412 (2002)
10. Joshi, A., Rowe, J.E., Zarges, C.: An immune-inspired algorithm for the set cover problem. In: Bartz-Beielstein, T., Branke, J., Filipič, B., Smith, J. (eds.) PPSN 2014. LNCS, vol. 8672, pp. 243–251. Springer, Heidelberg (2014)

11. Kim, J., Bentley, P.J.: Towards an artificial immune system for network intrusion detection: an investigation of clonal selection with a negative selection operator. In: CEC, vol. 2, pp. 1244–1252. IEEE Press (2002)
12. Laumanns, M., Thiele, L., Deb, K., Zitzler, E.: Combining convergence and diversity in evolutionary multiobjective optimization. Evol. Computat. **10**(3), 263–282 (2002)
13. Laumanns, M., Zitzler, E., Thiele, L.: On the effects of archiving, elitism, and density based selection in evolutionary multi-objective optimization. In: Zitzler, E., Deb, K., Thiele, L., Coello Coello, C.A., Corne, D.W. (eds.) EMO 2001. LNCS, vol. 1993, pp. 181–196. Springer, Heidelberg (2001)
14. Murphy, K.: Janeway's Immunobiology. Garland Science, New York (2011)
15. Schaffer, J.D.: Multiple objective optimization with vector evaluated genetic algorithms. In: ICGA, pp. 93–100. Lawrence Erlbaum Associates (1985)
16. Shah, R., Reed, P.: Comparative analysis of multiobjective evolutionary algorithms for random and correlated instances of multiobjective d-dimensional knapsack problems. Eur. J. Oper. Res. **211**(3), 466–479 (2011)
17. Sim, K., Hart, E., Paechter, B.: A lifelong learning hyper-heuristic method for bin packing. Evol. Comput. (2014, to appear)
18. Van Veldhuizen, D.A.: Multiobjective evolutionary algorithms: classifications, analyses, and new innovations. Technical report, DTIC Document (1999)
19. Zhang, Y., Meyer-Hermann, M., George, L.A., Figge, M.T., Khan, M., Goodall, M., Young, S.P., Reynolds, A., Falciani, F., Waisman, A., Notley, C.A., Ehrenstein, M.R., Kosco-Vilbois, M., Toellner, K.M.: Germinal center B cells govern their own fate via antibody feedback. J. Exp. Med. **210**(3), 457–464 (2013)
20. Zhou, A., Jin, Y., Zhang, Q., Sendhoff, B., Tsang, E.P.K.: Combining model-based and genetics-based offspring generation for multi-objective optimization using a convergence criterion. In: CEC, pp. 892–899. IEEE (2006)
21. Zitzler, E.: Evolutionary algorithms for multiobjective optimization: methods and applications. Ph.D. thesis, ETH Zurich, Switzerland (1999)
22. Zitzler, E., Laumanns, M., Thiele, L.: SPEA2: improving the strength pareto evolutionary algorithm. TIK report 103, Computer Engineering and Networks Laboratory (TIK), ETH Zurich (2001)
23. Zitzler, E., Thiele, L.: Multiobjective evolutionary algorithms: a comparative case study and the strength pareto approach. IEEE Trans. Evol. Comput. **3**(4), 257–271 (1999)

Mixing Network Extremal Optimization for Community Structure Detection

Mihai Suciu[1]([✉]), Rodica Ioana Lung[2], and Noémi Gaskó[1]

[1] Department of Computer Science, Babeş-Bolyai University,
Cluj Napoca, Romania
mihai-suciu@cs.ubbcluj.ro
http://csc.centre.ubbcluj.ro
[2] Faculty of Economics and Business Administration, Babeş-Bolyai University,
Cluj Napoca, Romania

Abstract. Mixing Network Extremal Optimization is a new algorithm designed to identify the community structure in networks by using a game theoretic approach and a network mixing mechanism as a diversity preserving method. Numerical experiments performed on synthetic and real networks illustrate the potential of the approach.

Keywords: Network structure · Extremal optimization · Nash equilibrium

1 Introduction

One of the most challenging problems in network analysis is the identification of the community structure, with applications in various fields such as politics, economics, biology, physics, etc. [1]. The main difficulty of this problem comes from the lack of a formal and a universal accepted definition for the concept of community.

Intuitively, a community is described as a group of nodes that are densely linked to each other and sparsely connected to the outside. The words "densely" and "sparsely" are difficult to transfer in a formal definition without introducing a threshold: how many inner links are enough to be considered "dense" and how many outside links are "sparse"? In [2], Radicchi defines the *strong community* as a group of nodes such that for each node the number of links inside the community is strictly greater than the number of links to outside nodes. He also defines the *weak community* as a group of nodes such that the total number of internal links is greater than the total number of external links. Other definitions are related to concepts from graph theory such as $LS - sets$ or k-core [3], but none of these encompass all features arising from the intuitive description of the community structure, especially when dealing with communities of different sizes and structures. In spite of that, a multitude of methods aiming to identify the community structure exist, validated mostly by means of numerical experiments.

Among these, some are based on defining the community structure as a partition maximizing a fitness function that is supposed to reflect all desired

© Springer International Publishing Switzerland 2015
G. Ochoa and F. Chicano (Eds.): EvoCOP 2015, LNCS 9026, pp. 126–137, 2015.
DOI: 10.1007/978-3-319-16468-7_11

properties of the structure. These optimization based approaches present the advantage that they can be used in combination with various heuristic algorithms. For example, heuristics attempting to maximize the modularity function [4] can be found in [5–8]. The modularity density [9], introduced as an alternative to modularity, was also extensively used [8,10–14].

Other widely used functions for evaluating the quality of a community are the *community score* defined in [15] and the *community fitness* used in [16]. The two measures have been used with uni-objective genetic algorithms [15] and within multiobjective approaches [17,18]. However, as they are all defined, none of these functions are capturing the community structure better than others; for example the limits of modularity have been pointed out it [19] while other fitness functions have been explored in [20].

Any method that aims to identify the community structure of a network by means of an optimization heuristic has to take into account the relevance and properties of the fitness function used: does it really reflect the community structure? - in the sense that an optimum value of the fitness function corresponds to a correct structure; and are there any local optima that can hinder the search process? This paper proposes a possible solution for this problem by combining two novel approaches: a game theoretic one that replaces the fitness function with a game setting and a network mixing method to avoid premature convergence during the search.

2   The Community Detection Game

The problem of detecting the network community structure can be converted into a mathematical game [21] in which nodes (as players) choose a community (strategy) trying to increase a corresponding payoff function. Then, the solution of such a game (Nash equilibrium) is a community structure such that no node can unilaterally improve its payoff by changing community. Thus, in equilibrium, we have a community structure in which each node belongs to the community that ensures best payoff given the communities chosen by all other nodes.

Formally, the game $\Gamma = (N, S, U)$ is described as a triplet consisting of:

- $N = \{1, \ldots, n\}$, the set of players: network nodes;
- $S = S_1 \times S_2 \times \ldots \times S_n$, the set of strategy profiles of the game; an element $s \in S$, $s = (c_1, \ldots, c_n)$ is a strategy of the game, with c_i denoting the community chosen by player i; for example, for a network with 4 nodes, a strategy profile $s = (1, 2, 2, 3)$ is a partition with three communities, with node 1 in the community 1, nodes 2 and 3 in community 2 and node 4 in community 3.
- $U = \{u_i\}_{i=\overline{1,n}}$, the payoff functions; $u_i : S \to \mathbb{R}$ represents the payoff of player i, $i = \overline{1, n}$.

A Nash equilibrium (NE) of a game is a strategy profile s^* such that the inequality $u_i(s_i, s^*_{-i}) \le u_i(s^*)$ holds for all $s_i \in S_i$ and $\forall i \in N$, where $(s_i, s^*_{-i}) = (s^*_1, \ldots, s^*_{i-1}, s_i, s^*_{i+1}, \ldots, s^*_n)$.

A NE for game Γ is a partition $s^* = (c_1^*, c_2^*, \ldots c_n^*)$, where c_i^* represents the community of player i, such that given any other partition $s = (c_1, c_2, \ldots, c_n)$ and any node/player i we have $u_i(s_i, s_{-i}^*) \leq u_i(s_i^*)$ i.e. no node can improve its payoff by changing community while all others maintain theirs.

Payoff functions. The *node* payoff is computed using the fitness of the node defined in [16], as the "contribution" of the node to the fitness of the community:

$$u_i(s) = f(C_i) - f(C_i \backslash \{i\}),$$

where C_i represents the community of player i, $C_i \backslash \{i\}$ is the same community but without node i, and f is the community fitness

$$f(C) = \frac{k_{in}(C)}{(k_{in}(C) + k_{out}(C))^{\alpha}}, \tag{1}$$

with:

- $k_{in}(C)$ - double of the number of internal links in the community;
- $k_{out}(C)$ - number of external links of the community;
- α - a parameter controlling the size of the community.

Intuitively, the NE of this game is a network partition such that each node belongs to the community to which it contributes the most, given that all other nodes are fixed in their communities.

Computing Nash Equilibria. The NEs of a game can be computed by evolutionary computation means by using the Nash ascendancy relation as introduced in [22] and first adapted for the community detection game in [21]. The Nash ascendancy relation provides a method to compare two strategy profiles by using a relative quality indicator $k(s, q), s, q \in S$; for game Γ this would denote the number of nodes from s that improve their payoff by individually switching their community from s_i to q_i, with $s = (s_1, \ldots, s_n)$ and $q = (q_1, \ldots, q_n)$:

$$k(s, q) = card\{i \in N | u_i(q_i, s_{-i}) > u_i(s), q_i \neq s_i\}.$$

We say that s Nash ascends q, or that s is better than q in Nash sense, if the inequality $k(s, q) < k(q, s)$ holds. If s^* is a NE, there does not exist any $q \in S$ such that q Nash ascends s. The converse is also true, i.e. if for strategy s there does not exist any $q \in S, q \neq s$ such that q Nash ascends s then s is a NE of the game [22].

The main application of this result is that the Nash ascendancy relation can be embedded into an optimization heuristic to guide its search towards the NEs of a game instead of the optimum values of a function. The efficiency of this approach has been tested on Cournot large games [23,24], with best results obtained when using an algorithm based on Extremal optimization [25].

3   Mixed Network Extremal Optimization

The Mixed Network Extremal Optimization (MNEO) proposed in this paper is based on the Nash Extremal Optimization for the Dynamic Community Detection [21], adapted to perform in a static environment with the use of a network mixing mechanism and a new Extremal Optimization feature to improve its performance.

MNEO populations. MNEO evolves two equal sized populations of individuals encoding network partitions: a current population P and an archive A that preserves the best solutions found so far by each individual in P. Let M be the size of the two populations. Paired individuals from the two populations follow the rules of EO: individual P_j explores the search space; the best value found by P_j is preserved in $A_j, j = \overline{1, M}$.

Encoding. Individuals are encoded as vectors of integers of size n equal to the number of nodes in the network $P_j = (p_{j_1}, \ldots, p_{j_n}), j = \overline{1, M}$; for node i, p_{j_i} represents the community of node i for individual j. This representation matches the form of the strategy profiles of game Γ, as MNEO evolves strategy profiles of Γ in order to compute game equilibria.

Number of communities. Each individual searches for a fixed number of communities, which is set within a minimum possible value, c_{min}, and a maximum one, c_{max}. c_{min} and c_{max} are algorithm parameters. There are approximatively equal numbers of individuals searching for a given number of communities (depending on the size of the population).

Extremal optimization. Individuals in the current population P explore the search space by the rules of an EO-based iteration with the following modification: for each individual P_j, the k nodes having the worst fitnesses are randomly assigned to different communities. If the newly obtained partition P'_j Nash ascends the corresponding archive member A_j, it will replace it. If not, the search continues next iteration from P'_j. There is no communication among individuals in P (or in A), the only information exchange takes place between P_j and $A_j, j = \overline{1, M}$. The outline of the EO iteration is presented in Algorithm 1.

Within this variant of EO, two mechanisms ensure the diversity of the search: the fact that P'_j replaces P_j unconditionally (P_j replaces A_j only if it Nash ascends it); and simultaneously changing k nodes randomly. The first one is specific to standard EO, while the latter has to be discussed: large values of k lead to a better exploration of the space, while smaller values lead to a better exploitation. Obviously, very high values of k may hinder the search by making it totally random, while very small values may induce premature convergence.

To avoid either situations the value of k is adapted during the search by decreasing it starting from approximatively 10 % of the number of nodes in the first iteration in order to better explore the space at the beginning of the search, to 1 towards the last ones to enhance the exploitation capabilities at the end

---

Algorithm 1. Extremal Optimization iteration.
Input: Current population P, archive A;
Parameters: number of nodes to be changed k;
Archive member A_j, $j = \overline{1, M}$, preserves the best solution found so far by P_j.

1: **for** each individual P_j in P **do**
2:    Select the k worst nodes from P_j;
3:    Randomly assign another community to each selected node - create offspring P'_j;
4:    **if** (P'_j Nash ascends A_j) **then**
5:       set $A_j := P_j$;
6:    **end if**
7:    set $P_j = P'_j$
8: **end for**

---

of the search. At current generation $NrGen, k$ is computed using the following formula:

$$k_{ngen} = \max\left\{1, \left[\frac{1}{10} \cdot N \cdot (N-2)^{-\frac{NrGen}{MaxGen}}\right]\right\},\tag{2}$$

where $[\cdot]$ represents the integer part, N the number of network nodes, and $MaxGen$ the maximum number of generations/iterations allowed.

Mixing the network. The main diversity preservation mechanism within MNEO consists on periodically rewiring the network, for small periods of time, in order to escape local optima and advance the search. For a number λ of iterations MNEO actually performs the search on a modified network; after that the original structure is restored and the search continues for Λ generations. Thus, the search alternates for Λ iterations on the original network and for λ iterations on the modified network.

The search performed on the modified network triggers a shift in the population. We can assume that small shifts in the network will cause small shifts in the population that will alter individuals just enough to ensure a better exploitation of the search space.

At the moment the network is modified (either altered or restored) both populations are evaluated and individuals from archive A are randomly re-initialized in order to allow the exploration; i.e. to be easily replaced by new solutions ensuring search diversity. Without this reinitialization local optimal from A cannot be replaced and the mixing mechanism looses its purpose.

The mixing mechanism used by MNEO, proposed in [26] and designed to preserve the node degrees in the network, randomly chooses a pair of links, erases them and re-connects the same nodes in a different manner. Links are selected randomly with a probability ρ controlling the magnitude of change in the network.

Outline of MNEO. MNEO runs a maximum number $MaxGen$ of EO iterations, alternating the search on the original network for Λ iterations with the search on the modified network for λ iterations. Algorithm 2 presents the outline of MNEO.

Algorithm 2. Outline of **Mixed Network Extremal Optimization.**

1: Randomly initialize all individuals in P and A;
2: Evaluate all individuals in P and A;
3: **for** $NrGen$ from 0 to $MaxGen$ **do**
4:    Set k_{ngen} according to eq. (2) - number of nodes with lowest fitness to be changed within an EO iteration;
5:    Run an EO iteration (alg. 1);
6:    **if** Λ iterations were performed on the original network **then**
7:       Mix network with probability ρ;
8:       Randomly re-initialize population A;
9:       Evaluate all individuals in P and A;
10:    **end if**
11:    **if** Network is mixed and λ iterations were performed **then**
12:       Restore network to original structure;
13:       Randomly re-initialize population A;
14:       Evaluate all individuals in P and A;
15:    **end if**
16: **end for**
17: **Output:** individual from A having the best community score [15];

Output. The individual with best fitness: any function measuring the quality of the cover may be used. However, this value is not actually used during the search process, only to select the output of the algorithm from the final population.

Parameters. MNEO uses the following parameters:

– Population size;
– Maximum number of iterations $MaxGen$;
– ρ - network mixing probability;
– Λ and λ - number of iterations MNEO performs the search on the original network and the modified one, respectively;
– c_{min} and c_{max} - minimum and maximum number of communities searched for.

4 Numerical Results and Discussions

4.1 Experimental Setting

The performance of MNEO is tested on the GN [4] and LFR [27] benchmarks[1]. The GN networks have 128 nodes with degree 16 and 4 communities (each community has 32 nodes). We used 8 sets of 30 GN networks each, with $z_{out} \in \{1, 2, \ldots, 8\}$, where z_{out} represents the number of edges linking nodes to other communities and $16 = z_t = z_{in} + z_{out}$. We used 6 sets of 30 LFR networks with

---

[1] Generated with the code available at http://sites.google.com/site/andrealancichinetti/files, downloaded on March 2014.

$\mu \in \{0.1, 0.2, \ldots, 0.6\}$, where μ represents the mixing parameter and is the ratio between z_{out} and z_t. The parameters used to generate the LFR networks are: average vertex degree 20, maximum vertex degree 50, number of nodes 128, and community size $[10, 50]$ for the *small* set and $[20, 100]$ for the *big* set.

Four real-world networks having known community structures are also considered: Zachary's karate club, bottle-nose dolphins, Krebs political books, and the American college football.

The Normalized Mutual Information (NMI) indicator is used [16] to evaluate the quality of the solutions and to compare results with other methods. A NMI value of 1 (maximum possible) indicates that the correct cover has been identified.

4.2   Setting Parameters

In order to study the effects different parameter settings have on MNEO, several settings have been tested using the GN and LFR benchmarks, with the following values:

- population size $M \in \{20, 50, 100\}$;
- number of iterations the network is mixed $\lambda \in \{10, 15, 20, 50, 100\}$;
- number of iterations the network is in original state $\Lambda \in \{10, 20, 30, 50, 100\}$;
- mixing probability $\rho \in \{0.001, 0.01, 0.05, 0.1\}$.

The 18 settings tested in order to asses if any of the parameters may have a special impact on the results (by keeping the others unchanged) can be grouped in four: settings I-III report for different population sizes; IV-VII report for different values of λ; VIII-XI report for different values of Λ; and XII-XV for different values of ρ.

The effect of the network mixing mechanism has been studied by disabling the mixing feature of MNEO (equivalent to $\rho = 0$).

Other parameters: minimum and maximum number of communities c_{min} and c_{max} were set 2 and 8, respectively. The α parameter in equation (1) is set to 1 as its effect is not in the scope of this study.

All comparisons are performed by using a Wilcoxon sum-rank test.

Results. For the GN and LFR networks with well defined community structures (GN $z_{out} = 1, \ldots, 6$ and LFR *small* and *big* with $\mu = 0.1, 0.2, 0.3$), MNEO computed the correct cover for all networks and all parameter settings, with average NMI=1, indicating the efficiency of the game theoretic approach by itself.

For GN $z_{out} = 7$, MNEO found the correct cover with NMI=1 in all runs, significantly different and better than the no-mix variant (according to a Wilcoxon sum-rank test with 5 % confidence level). Up to this point, it can be inferred from the results that when the community structure is well defined, the different parameter settings considered do not influence results significantly.

Other results are depicted in Fig. 1. The following conclusions can be drawn regarding to the effects of the parameter settings on MNEO:

- I-III: the size of the population influences the results of the MNEO search and of the no-mix variant; mean NMI color bars indicate that best results may be obtained for $M = 100$;
- IV-VII: number λ of iterations the search takes place almost never leads to statistically significant different results;
- VIII-XI: number Λ of iterations MNEO searches on the original network: if it is too high the results are significantly different in some cases, but worse than for lower values; it is interesting to notice that for the smallest value (setting VIII, $\Lambda = 10$) sometimes results are significantly better than for other values of Λ;
- XII-XV: mixing probability ρ influences the search with best results obtained for $\rho = 0.05$ and worst for $\rho = 0.1$ and 0.001.
- in most cases the results obtained by MNEO are significantly better than the no-mix variant; moreover, there is only one instance in which MNEO results are worst (XV) for $\rho = 0.1$ suggesting that the magnitude induced by this value is too high, disrupting the search.

4.3 Comparisons with Other Methods

The results of MNEO for the tested benchmarks are compared with four algorithms: OSLOM [28], Infomap [29], Modularity optimization (ModOpt) [30], and the Louvain method [31][2]. For the synthetic networks all algorithms are run once on each of the 30 network realizations; for the real networks 30 runs are performed; a Wilcoxon sum-rank test is used to determine if there are statistical differences between results obtained with MNEO and those of the other methods. Figure 2 presents mean NMI values with error bars for all five methods.

Results show that for the synthetic networks, when the community structure is well defined ($z_{out} = 1, 2, 3, 4, 5$ for GN and $\mu = 0.1, 0.2, 0.3$ for LFR small set), all algorithms find the true community structure of the network. When the structure weakens ($z_{out} = 6, 7, 8$ for GN and $\mu = 0.4, 0.5, 0.6$ for LFR small and big sets), MNEO results are significantly better than all the others.

For the four real-world networks, MNEO provides significantly better results than those of the other algorithms (Wilcoxon $p < 0.05$), except for the Karate network where the difference between results obtained by MNEO and OSLOM is not significant ($p = 0.1961$). Boxplots of the results are presented in Fig. 3. While further analysis is required to assess the efficiency of MNEO, the results presented in this paper places it among promising methods for computing network community structure.

---

[2] For the algorithms we use the code and parameter setting available at https://sites.google.com/site/andrealancichinetti/software, downloaded on March 2014.

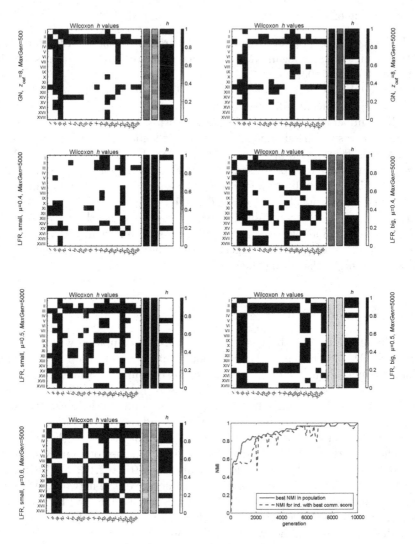

Fig. 1. For each network (left/right labels): a matrix containing Wilcoxon logical h values indicating weather there is a statistical significant difference between different parameter settings; two color bars for the Average NMI obtained by MNEO and the corresponding no-mix variant, and a color bar with the Wilcoxon h values for the differences between the two, for each setting. Black boxes indicate significant differences (Wilcoxon h values are 1, and the null hypothesis is rejected) and white boxes indicate that the null hypothesis of equalities between medians cannot be rejected and $h = 0$. The last figure illustrates the evolution of NMI on a GN network with $z_{out} = 8$ averaged over 10 runs. The best NMI in the population and the NMI of the individual with the best community score are plotted.

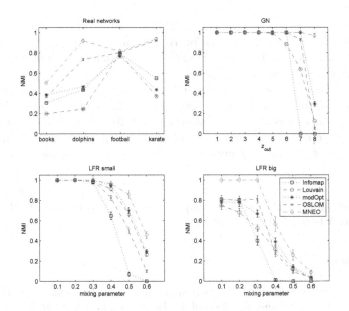

Fig. 2. Comparisons with other methods - mean NMI (with error bars) values obtained in 30 runs (30 instances for each GN and LFR networks and 30 independent runs for the real-world networks).

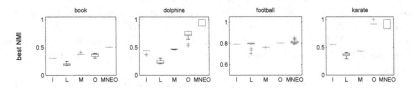

Fig. 3. Boxplots of results obtained on the real networks with: I (Infomap), L (Louvain), M (ModOpt), O (Oslom), and MNEO.

5   Conclusions

MNEO is a heuristic designed to identify network community structures. MNEO converts the problem into a mathematical game in which nodes act as players choosing communities that maximize their payoff. The search for the Nash equilibrium of the game is performed using an adapted extremal optimization algorithm. To avoid premature convergence, the network is periodically shuffled (mixed) by adding and removing links with a given probability. Shifts in the network trigger corresponding shifts in the population, allowing it to escape local optima.

Numerical experiments illustrate the behavior of MNEO on small (128 nodes) synthetic networks (GN and LFR) and on four real-world networks. MNEO provides results significantly better or at least as good as the other methods considered for the tested benchmarks.

Although results are promising, MNEO is a computationally expensive method, as the Nash ascendancy relation requires at least $2k_{ngen}$ payoff function evaluation on each call. However, MNEO may be useful in refining results provided by other, faster methods.

Further work consists on studying the behavior of MNEO on larger networks and different network mixing mechanisms. The game theoretic design can be improved also by considering different payoff functions. The study of other equilibria concepts, either refinements of the Nash equilibrium or totally different ones, such as the Berge-Zhukovskii equilibrium, may also lead to interesting results.

Acknowledgment. This work was supported by the project "OPEN-RES (PN-II-PC-CA-2011-3.1-0682 212/2.07.2012)."

References

1. Fortunato, S.: Community detection in graphs. Phys. Rep. **486**, 75–174 (2010)
2. Radicchi, F., Castellano, C., Cecconi, F., Loreto, V., Parisi, D.: Defining and identifying communities in networks. Proc. Natl. Acad. Sci. U.S.A **101**(9), 2658–2663 (2004)
3. Wasserman, S., Faust, K.: Social Network Analysis: Methods and Applications, vol. 8. Cambridge University Press, Cambridge (1994)
4. Girvan, M., Newman, M.E.J.: Community structure in social and biological networks. Proc. Natl. Acad. Sci. **99**(12), 7821–7826 (2002)
5. Nascimento, M.C., Pitsoulis, L.: Community detection by modularity maximization using GRASP with path relinking. Comput. Oper. Res. **40**(12), 3121–3131 (2013)
6. Shang, R., Bai, J., Jiao, L., Jin, C.: Community detection based on modularity and an improved genetic algorithm. Phys. A: Stat. Mech. Appl. **392**(5), 1215–1231 (2013)
7. Honghao, C., Zuren, F., Zhigang, R.: Community detection using ant colony optimization. In: 2013 IEEE Congress on Evolutionary Computation (CEC), pp. 3072–3078 (2013)
8. Shi, C., Yan, Z., Cai, Y., Wu, B.: Multi-objective community detection in complex networks. Appl. Soft Comput. **12**(2), 850–859 (2012)
9. Li, Z., Zhang, S., Wang, R.S., Zhang, X.S., Chen, L.: Quantitative function for community detection. Phys. Rev. E **77**, 036109 (2008)
10. Gong, M., Fu, B., Jiao, L., Du, H.: Memetic algorithm for community detection in networks. Phys. Rev. E **84**, 056101 (2011)
11. Jiang, J.Q., McQuay, L.J.: Modularity functions maximization with nonnegative relaxation facilitates community detection in networks. Phys. A: Stat. Mech. Appl. **391**(3), 854–865 (2012)
12. Gong, M., Ma, L., Zhang, Q., Jiao, L.: Community detection in networks by using multiobjective evolutionary algorithm with decomposition. Phys. A: Stat. Mech. Appl. **391**(15), 4050–4060 (2012)
13. Gong, M., Cai, Q., Chen, X., Ma, L.: Complex network clustering by multiobjective discrete particle swarm optimization based on decomposition. IEEE Trans. Evol. Comput. **18**(99), 82–97 (2013)

14. Angelini, L., Boccaletti, S., Marinazzo, D., Pellicoro, M., Stramaglia, S.: Identification of network modules by optimization of ratio association. Chaos: An Interdisc. J. Nonlinear Sci. **17**(2), 023114 (2007)
15. Pizzuti, C.: GA-Net: a genetic algorithm for community detection in social networks. In: Rudolph, G., Jansen, T., Lucas, S., Poloni, C., Beume, N. (eds.) PPSN 2008. LNCS, vol. 5199, pp. 1081–1090. Springer, Heidelberg (2008)
16. Lancichinetti, A., Fortunato, S., Kertész, J.: Detecting the overlapping and hierarchical community structure in complex networks. New J. Phys. **11**(3), 033015 (2009)
17. Amiri, B., Hossain, L., Crawford, J.W., Wigand, R.T.: Community detection in complex networks: multi-objective enhanced firefly algorithm. Knowl.-Based Syst. **46**, 1–11 (2013)
18. Pizzuti, C.: A multiobjective genetic algorithm to find communities in complex networks. IEEE Trans. Evol. Comput. **16**(3), 418–430 (2012)
19. Lancichinetti, A., Fortunato, S.: Limits of modularity maximization in community detection. Phys. Rev. E **84**, 066122 (2011)
20. Chira, C., Gog, A., Iclanzan, D.: Evolutionary detection of community structures in complex networks: A new fitness function. In: IEEE Congress on Evolutionary Computation (CEC), 2012, pp. 1–8. IEEE (2012)
21. Lung, R.I., Chira, C., Andreica, A.: Game theory and extremal optimization for community detection in complex dynamic networks. PLOS ONE **9**(2), e86891 (2014)
22. Lung, R.I., Dumitrescu, D.: Computing nash equilibria by means of evolutionary computation. Int. J. Comput. Commun. Control **III**(suppl.issue), 364–368 (2008)
23. Lung, R.I., Mihoc, T.D., Dumitrescu, D.: Nash equilibria detection for multi-player games. In: IEEE Congress on Evolutionary Computation, 1–5 (2010)
24. Lung, R.I., Mihoc, T.D., Dumitrescu, D.: Nash extremal optimization and large cournot games. In: Pelta, D.A., Krasnogor, N., Dumitrescu, D., Chira, C., Lung, R. (eds.) NICSO 2011. SCI, vol. 387, pp. 195–203. Springer, Heidelberg (2011)
25. Boettcher, S., Percus, A.G.: Optimization with extremal dynamics. Phys. Rev. Lett. **86**, 5211–5214 (2001)
26. Maslov, S., Sneppen, K.: Specificity and stability in topology of protein networks. Sci. **296**, 910–913 (2002)
27. Lancichinetti, A., Fortunato, S.: Benchmarks for testing community detection algorithms on directed and weighted graphs with overlapping communities. Phys. Rev. E **80**, 016118 (2009)
28. Lancichinetti, A., Radicchi, F., Ramasco, J.J., Fortunato, S.: Finding statistically significant communities in networks. PloS one **6**(4), e18961 (2011)
29. Rosvall, M., Bergstrom, C.T.: Maps of random walks on complex networks reveal community structure. Proc. Natl. Acad. Sci. **105**(4), 1118–1123 (2008)
30. Sales-Pardo, M., Guimer, R., Moreira, A., Nunes Amaral, L.: Extracting the hierarchical organization of complex systems. Proc. Natl. Acad. Sci. U.S.A. **104**(39), 15224–15229 (2007)
31. Blondel, V.D., Guillaume, J.L., Lambiotte, R., Lefebvre, E.: Fast unfolding of communities in large networks. J. Stat. Mech.: Theory Exp. **2008**(10), P10008 (2008)

Multi-start Iterated Local Search for the Mixed Fleet Vehicle Routing Problem with Heterogenous Electric Vehicles

Ons Sassi[1]([✉]), W. Ramdane Cherif-Khettaf[1], and Ammar Oulamara[2]

[1] University of Lorraine - LORIA, Nancy, France
{ons.sassi,wahiba.ramdane}@loria.fr
[2] University of Lorraine, Ile de Saulcy, Metz, France
ammar.oulamara@loria.fr

Abstract. This paper deals with a real world application that consists in the vehicle routing problem with mixed fleet of conventional and heterogenous electric vehicles including new constraints, denoted VRP-HFCC. This problem is defined by a set of customers that have to be served by a mixed fleet of vehicles composed of heterogenous fleet of Electric Vehicles (EVs) with distinct battery capacities and operating costs, and a set of identical Conventional Vehicles (CVs). The EVs could be charged during their trips in the available charging stations, which offer charging with a given technology of chargers and time dependent charging costs. Charging stations are also subject to operating time windows constraints. EVs are subject to the compatibility constraints with the available charging technologies and they could be partially charged. Intermittent charging at the depot is also allowed provided that constraints related to the electricity grid are satisfied. The objective is to minimize the number of employed vehicles and to minimize the total travel and charging costs. The developed multi-start algorithm is based on the Iterated Local Search metaheuristic which uses a Large Neighborhood Search with two different insertion strategies in the Local Search procedure. Different implementation schemes of the proposed method are tested on a set of real data instances with up to 550 customers as well as on generalized benchmark instances.

Keywords: Vehicle routing problem · Electric vehicle charging · Metaheuristics · Iterated Local Search · Large Neighborhood Search · Optimization

1 Introduction

To tackle environmental challenges, investing in more environmentally friendly modes of transportation such as the ridesharing service [1] and Electric Vehicles (EVs) is becoming a necessity. In fact, EVs may provide a clean alternative to the conventional vehicles (CVs). However, electric vehicle industry is still facing many weaknesses related to the battery management and the charging infrastructure.

© Springer International Publishing Switzerland 2015
G. Ochoa and F. Chicano (Eds.): EvoCOP 2015, LNCS 9026, pp. 138–149, 2015.
DOI: 10.1007/978-3-319-16468-7_12

In this paper, we address the new vehicle routing problem with mixed fleet of conventional and heterogenous electric vehicles, in which a set of customers have to be served by a fleet of CVs and EVs operating with plug-in batteries. EVs need to be charged in charging stations during the trips in order to serve all customers. This real-world problem was addressed in the framework of the French national R&D project Infini Drive, led by La Poste Group, ERDF (French Public Electricity Distribution Network Manager) and seven other companies and research laboratories. Furthermore, this study follows on from the work presented in [13] where exact and heuristic methods were presented to solve the joint EV scheduling and charging problem. Within this study, we extend this problem to the case where the routes have to be constructed and assigned to the available vehicles with the objective of minimizing the overall routing and charging costs. In the case where EVs routes are not already constructed, we refer to the Electric Vehicle Routing Problem which is an extension of the more general Vehicle Routing Problem (see for example [16]).

The problem of energy-optimal routing is addressed in [2]. In [4], the authors formulate the Green Vehicle Routing Problem (GVRP) as a Mixed Integer Linear Program (MIP). Two constructive heuristics are developed to solve this problem. An overview of the GVRP is given in [9]. Schneider et al. [14] combine a Vehicle Routing Problem with the possibility of refueling a vehicle at a station along the route. They introduce the Electric Vehicle Routing Problem with Time Windows and Recharging Stations (E-VRPTW). E-VRPTW aims at minimizing the number of employed vehicles and total traveled distance. We are also aware of more recent studies that were conducted simultaneously with our work. In [6], the Electric Vehicle Routing Problem with Time Windows and Mixed Fleet to optimize the routing of a mixed fleet of EVs and CVs is addressed. On each visit to a recharging station, EVs are recharged to their maximum battery capacity with a constant recharging rate. To solve this problem, an Adaptive Large Neighborhood Search algorithm that is enhanced by a local search for intensification is proposed. Almost the same problem is addressed in [8]. The only difference here is the fact of considering a heterogenous fleet of vehicles that differ in their transport capacity, battery size and acquisition cost. An Adaptive Large Neighbourhood Search with an embedded local search and labelling procedure for intensification is developed. In [5], the authors present a variation of the electric vehicle routing problem in which different charging technologies are considered and partial EV charging is allowed. This problem is the closest to our problem in the sense that we consider different charging technologies and partial EV charging. However, several major differences have to be outlined. Firstly, we consider a mixed fleet composed of CVs and heterogenous EVs. Secondly, the costs of charging at the depot and at the charging stations are assumed to be time dependent. Moreover, the charging stations are subject to operating time windows constraints and charging at the depot is subject to the grid's maximum capacity constraints. Besides, EVs are not necessarily compatible with all charging technologies.

In short, we differ from all the above-mentioned studies in that we consider, within the same study, a mixed fleet composed of CVs and heterogenous EVs, different types of charging stations and different time-dependent charging costs. Moreover, EV charging at the depot could be intermittent and is subject to real-life constraints such as the maximum grid capacity constraint. We also consider that not all EV are compatible with fast charging technologies and that partial charging is allowed. Our objective function is also different. In fact, we aim at minimizing total operating and charging costs involved with the use of a mixed fleet. Our overall objective is to provide enhanced optimization methods for EV charging and routing that are relevant to the described constraints.

To solve the VRP-HFCC, we develop a multi-start Iterated Local Search metaheuristic that uses a Local search based on a Large Neighborhood Search (LNS) with two different insertion strategies. The LNS was first proposed by Shaw [15], and later adapted by Pisinger and Ropke [12]. The developed algorithm is tested on a set of real data instances as well as on generalized benchmark instances following different implementation schemes. The remainder of the paper is organized as follows. In Sect. 2, we introduce the notation in detail. In Sect. 3, our solving approach is presented. Section 4 summarizes the computational results. Concluding remarks are given in Sect. 5.

2   Problem Description and Notation

We define the VRP-HFCC on a complete, directed graph $G = (V', A)$. V' denotes the set of vertices composed of the set V of n customers, the set F of charging stations $F = \{1, \ldots, f\}$ and the set $D = \{1, \ldots, |D|\}$ of chargers at the depot. The set of arcs is denoted by $A = \{(i, j) \mid i, j \in V', i \neq j\}$. The depot is denoted by either 0 or $n + 1$ depending if it is the initial or terminal node of a route.

Our optimization time horizon $[0, T]$, which represents typically a day, is divided into T equidistant time periods, $t = 1, \ldots, T$, each of length δ, where t represents the time interval $[t - 1, t]$. We define the night interval $[0, T_0] \subset [0, T]$ during which charging at the depot with Level 1 chargers could be performed. Moreover, no customer has to be served during the night period. We define the service interval $[T_0, T] \subset [0, T]$ during which all customers have to be served and the EVs could be charged in the different charging stations as well as in the depot using the available chargers. A nonnegative demand q_i is associated with each customer $i \in V$, this represents the quantity of goods that will be delivered to this customer. With each customer we also associate a service time s_i. Each arc $(i, j) \in A$ is defined by a distance $d_{i,j}$ and a nonnegative travel time $t_{i,j}$ required to travel $d_{i,j}$. When an arc (i, j) is traveled by an EV, it consumes an amount of energy $e_{i,j}$ equal to $r \times d_{ij}$, where r denotes a constant energy consumption rate.

Each charging station $f \in F$ can deliver a maximum charging power p_f (kW) and proposes a time dependent charging cost $c_{f,t}, \forall t = T_0, \ldots, T$; which represents the charging cost during the time period t, expressed in (euros/kWh). The chargers in charging station f are available during the time window $[a_f, b_f]$.

Accordingly, the EV must wait if it arrives at charging station f before time a_f. Note that, within this study, we consider that the charging stations could propose three different charging technologies: (i) Level 1 charger which is the slowest charging level that provides charging with a power of 3.7 kW; (ii) Level 2 charger offers charging with a power of 22 kW and (iii) Level 3 charger which is the fastest charging level that delivers a power of 53 kW.

We consider a set $M_{EV} = \{1,\ldots,m_{EV}\}$ of EVs and a set $M_{CV} = \{m_{EV} + 1,\ldots,m_{EV} + m_{CV}\}$ of Combustion Engine Vehicles (CVs), needed to serve all customers. Each EV k operates with a battery characterized by its nominal capacity of embedded energy CE_k(kWh) and its State of Charge (SoC_k^0) at time $t = 0$ expressed as a ratio of the available amount of energy and CE_k (0 = empty; 1 = full). At low and high SoC's values, the battery tends to degrade faster [3,11]. In order to improve its lifetime after repeated use and to respect the security issues, at each time t, SoC_k^t should be in the interval $[SoC_k^{Min}, SoC_k^{Max}]$, where SoC_k^{Min} and SoC_k^{Max} are the minimal and maximal allowable values of SoC, respectively. Each EV (CV) is characterized by a maximum capacity Q^{EV} (Q^{CV}) which represents the maximum quantity of goods that could be transported by the vehicle. Denote by FC^{EV} (FC^{CV}) (euros/day) the fixed costs related to EVs (CVs). Denote by OC_k^{EV} (OC^{CV}) the operating costs (euros/km) related to the maintenance of EV k (CV), accidents, etc. Thus, if an arc (i,j) is traveled by an EV k (CV), this has an operating cost denoted by $cost_{i,j,k}^{EV}$ ($cost_{i,j}^{CV}$) and is computed as: $cost_{i,j,k}^{EV} = d_{i,j} \times OC_k^{EV}$ ($cost_{i,j}^{CV} = d_{i,j} \times OC^{CV}$).

At the depot, a given number of Level 1 chargers are available to charge the EVs during the optimization horizon $[0,T]$ and a predefined number of Level 2 chargers are available to charge the vehicles only during the service time $[T_0,T]$. At each time period t, each charger at the depot can apply on EV k a charging power $p_{kt} \in [p^{Min}, p^{Max}]$ where p^{Min} and p^{Max} are the minimal and maximal powers that can be delivered by the charger, respectively. Thus, an EV charged with a power p_{kt} during the time period t retrieves a total amount of energy equal to $\delta \times p_{kt}$(kWh). We denote by GP_t the electricity grid capacity available for EV charging at time t; i.e., at each time period t, the total grid power available to charge all EVs is limited to GP_t. Let c_t' be the energy cost during t.

Each customer $i \in V$ should be visited, by either an electric or conventional vehicle, exactly once during $[T_0,T]$. Each charging station could be visited as many times as required or not at all. When charging is undertaken in a charging station f, it is assumed that only the required quantity of energy is injected into the EV battery. Thus, EVs could be partially charged.

Since we consider many charging technologies (slow and fast charging), we should also consider the fact that not all EVs technologies are compatible with fast charging. Thus, when we plan the charging of an EV, only the charging stations proposing compatible charging technologies should be considered. A feasible solution to our problem is composed of a set of feasible routes assigned to adequate vehicles and a feasible EVs charging planning. A feasible route is a sequence of nodes that satisfies the following set of constraints:

– Each route must start and end at the depot;
– the overall amount of goods delivered along the route, given by the sum of the demands q_i for each visited customer, must not exceed the vehicle capacity (Q^{EV} or Q^{CV});
– the total duration of each route, calculated as the sum of all travel durations required to visit a set of customers, the time required to charge the vehicle during the interval $[T_0, T]$, the service time of each customer and, eventually, the waiting time of the EV if it arrives at a charging station before its opening time, could not exceed $T - T_0$;
– no more than m_{EV} EVs and m_{CV} CVs are used;
– each customer should be visited between T_0 and T;
– the following charging constraints are satisfied:
  • the charging level of the battery of each EV k must always be in the interval $[SoC_k^{Min}, SoC_k^{Max}]$;
  • when charging is undertaken, each EV should be charged with a compatible charging technology;
  • when EVs are charged at the depot, the total power used to charge them does not exceed the grid's maximum capacity and the minimum and the maximum powers of chargers should be respected;
  • during the time interval $[0, T_0]$, EV charging at the depot could only be performed using the available Level 1 chargers;
  • at each charging station f, charging could only be undertaken during its operating time window $[a_f, b_f]$.

We seek to construct a minimum number of routes such that all customers are served, all EVs are optimally charged and the total cost of routing and charging is minimized. The objective function, measured in monetary units, consists in minimizing five costs: (i) the routing cost that depends on the number of kilometers traveled by each vehicle and the vehicle operating cost, (ii) the charging cost engendered by charging EVs in the charging stations during $[T_0, T]$, (iii) the cost of charging EVs at the depot during $[0, T_0]$, (iv) the vehicles total fixed cost and (v) the total cost engendered by the waiting time of the EVs if they arrive at a charging station before its opening time.

3   Iterated Local Search Meta-Heuristic

Since the considered problem is NP-hard and in order to solve large instances of the VRP-HFCC, we develop an Iterated Local Search (ILS) metaheuristic. The main steps of ILS were first proposed by H.R. Lourenço et al. in [10]. The developed multi-start ILS algorithm uses five procedures: (i) generation of an initial solution; (ii) a Local Search which improves the solution initially obtained; (iii) a Perturbation Mechanism that generates a new starting point through perturbation of the solution returned by the Local Search; (iv) an Acceptance Criterion that specifies if the solution should be accepted or not and (v) a Stopping Criterion that specifies when the ILS procedure should stops. The multi-start heuristic executes a given number of iterations, where at each iteration a new starting solution is generated by a constructive heuristic. The generated initial solution is then improved using a LNS with two different insertions strategies in the Local Search phase combined with a perturbation phase. The next subsections describe the different ILS procedures.

3.1 Initial Solution Generation

An initial feasible solution is generated with a Charging Routing Heuristic (CRH). The CRH consists of two steps. In the first step, the heuristic generates a feasible charging scheme for EVs at the depot during $[0, T_0]$. It starts by sorting the time periods according to the ascending order of electricity costs. Let T_{sorted} be the sorted table of all time periods in $[0, T_0]$. With each electric vehicle k, we associate a priority $priority_t^k$ that translates the fact that EV k has higher priority to charging than the other available EVs during the time period t. The heuristic selects the first available time period in T_{sorted} as well as the EV with the highest priority and charges it with the minimal possible charging power between: (i) the maximal power of chargers, (ii) the grid's capacity that is still available, and (iii) the maximum power that will completely full the vehicle's battery. The CRH selects then a new different EV. This procedure is repeated until no possible charging could be undertaken. In the second step, the CRH generates a routing and charging schedule for all EVs during the service interval. Algorithm 1 gives more details about the CRH heuristic.

3.2 Local Search Procedure

The Local Search procedure is based on a LNS (LNS-LS). To the best of our knowledge, no previous study has been conducted to solve routing problems using an ILS with a LNS-based local search.

The following parameters are useful in the LNS-LS:

- $Iter$: parameter that controls the size of the main loop of the algorithm.
- $IterLNS$: parameter that specifies the number of times the LNS should be repeated.
- $trial$: parameter that specifies the number of times the insertion procedure should be repeated in order to find the best improvement insertion
- Num: parameter that controls the size of the neighborhood list that will be used in the LNS.

The Local Search procedure restarts $Iter$ times and for each new best solution, it performs $IterLNS$ iterations of the following neighborhood ejection and injection strategy. A node j and a set of $Num-1$ additional nodes located the nearest possible to j (in terms of costs), are randomly selected (the selected neighbors may be in different routes). This neighborhood of Num nodes is then ejected from the solution. The ejected nodes are then re-inserted back into the partial solution using one of the two different insertion methods: (i) random insertion and (ii) insertion method with regret search. For each list of ejected nodes, the

---

Algorithm 1. Charging Routing Algorithm

1: **Input:** A graph $G = (V', A)$ and a set of $m_{EV} + m_{CV}$ empty routes
2: **Output:** A set of routes assigned to at most $m_{EV} + m_{CV}$ vehicles
3: **Step 1** : Generate a charging schedule for all EVs at the depot during $[0, T_0]$
4: **Step 2**
5: **while** the maximum number of routes is not yet reached AND there exists at least one customer that is not yet served **do**
6:    Select the EV k with the lowest priority at $t = T_0$ among all available EVs not yet assigned
7:    **while** the total route duration is less than $T - T_0$ and the total amount of goods delivered along the route is less than Q^{EV} **do**
8:       Sort the list of nodes randomly and let $V(i)$ be the set of all neighbors of node i not yet visited and that could be visited using the remaining battery energy of the current vehicle
9:       **if** $V(i)$ contains at least one customer and either the depot or a charging station $f \in V(j) \cap F(j)$ **then**
10:          select a node j from $V(i)$ such that $cost_{i,j,k}^{EV}$ is minimal
11:       **else if** $(V(j)$ is empty or it contains only customers or incompatible charging stations) AND (charging is possible) **then**
12:          the vehicle should get charged before visiting j, in that case insert the compatible charging station with the lowest cost while ensuring that this charging station will be available when the EV arrives at this station
13:       **else**
14:          Assign i to the CV having a sufficient capacity and engendering a minimum insertion cost
15:       **end if**
16:    **end while**
17: **end while**

---

insertion procedure is repeated *trial* times and, at the end, the ejected nodes are re-inserted in the route positions engendering the best improvement in the solution cost. If the solution becomes infeasible, we insert a new charger, having the lowest cost, in the route while ensuring that the constraints related to the compatibility of the charging stations with the EV as well as the station's operating time windows constraints are satisfied. If it is not possible to insert the ejected node in an already constructed route, a new route that contains this node and the depot may be created. In that case, the vehicle ownership cost is added to the total route cost.

When all customers have been re-inserted back into the solution using one of the two insertion methods, the new solution is compared with the original solution. If the resulting solution is better than the original solution, then the next iteration continues with the new solution. Otherwise, the next iteration continues with the original solution. After *Iter* runs, the best solution found during the search is reported.

In the following, we detail the insertion methods.

Algorithm 2. Local Search procedure

1: **for** $i = 0$ to $Iter$ **do**
2:    Let s' be the best generated solution
3:    **for** $j = 0$ to $IterIE$ **do**
4:       Eject a list of Num nodes
5:       **for** $k = 0$ to $trial$ **do**
6:          Insert the ejected nodes in the cheapest route positions following a given insertion strategy and let s'' be the obtained solution
7:          **if** $total\_route\_cost < best\_cost$ **then**
8:             $s''^* \leftarrow s''$
9:          **end if**
10:          Eject again the list of Num nodes
11:       **end for**
12:       **if** $cost_{s''^*} < cost_{s'}$ **then**
13:          $s' \leftarrow s''^*$
14:       **end if**
15:    **end for**
16: **end for**

Random Insertion Method. This method selects randomly a node among the list of ejected nodes and inserts it in the position that generates the minimal cost increase in the total solution cost. If the insertion of a customer in a given route position leads to a violation of the vehicle capacity or total time constraints, this route position will not be accepted. However, if the insertion of a customer in a given route position still satisfies the vehicle capacity and total time constraints but leads to a violation of the energy constraints (in the case where the EV needs more energy to serve this customer or the time planned for charging decreases since it depends on the opening time windows of the charging stations), this method tries to repair the solution by inserting chargers in the route while ensuring the compatibility between the EV and the chargers and satisfying the charging stations' operating time windows constraints. At each update of the routing and charging solution, the total solution cost is updated.

Insertion Method with Regret Search. The insertion method with regret search uses the same cheapest insertion method as the random insertion method, but allows previous insertions to be undone if this removal allows for a cheaper insertion of the current customer under consideration. This is similar to the notion of regret described in [7]. At each step, the cheapest next insertion and the maximum cost reduction caused by deleting a node (which is not one of the partial solution vertices participating in the insertion) from the current partial solution are compared. The LNS moves remain temporary and become final only when all ejected nodes are re-injected.

3.3   Perturbation Mechanism

The solution generated by the local search procedure is perturbed to avoid stopping at a local optimum. The Perturbation mechanism uses the LNS but it explores a larger neighborhood space than the one explored by the Local Search. The perturbation mechanism consists in the following steps:

– Eject a random list of $Num_{perturb}$ nodes such that $Num_{perturb} > Num$.
– Inject randomly the ejected nodes.

3.4   Acceptance Criterion

To escape from a current locally optimal solution, non improving-solutions could be accepted. Our acceptance criterion is based on the mechanism of accepting non-improving solutions used by the Record-to-Record algorithm. During the run of the ILS procedure, any solution is accepted if its objective value is lower than $(1 + Dev) \times Record$, where the $Record$ is the value of best solution obtained and Dev is a parameter. Initially, $Record$ is equal to the initial objective function. During the search process, $Record$ is updated with the objective value of the best solution so far.

Now, we have all sub-routines to describe the ILS method. In the following, Algorithm 3 describes the ILS algorithm in detail. The parameter $Max\_Restart$ specifies the maximum number of iterations to be executed starting from a new initial solution generated by the CRH heuristic. The parameter max_{impr} represents the maximum number of consecutive perturbations allowed without improvement of the current best solution.

4   Computational Experiments and Discussion

Our methods were implemented using C++. All experiments were carried out on an Intel Xeon E5620 2.4 GHz processor, with 8 GB RAM memory. We conducted numerical experiments on real data instances provided by a French company. Experiments were conducted on 9 real data instances. The number of nodes for the considered instances ranges between 300 and 550 and the number of charging stations ranges between 15 and 35.

The characteristics of the real instances are:

– $m_{EV} = 18$; $m_{CV} = 8$; $Q^{EV} = 3$; $Q^{CV} = 5$; $CE_k = 22 \; \forall k \in 1, \ldots, 9$; $CE_k = 16$
   $\forall k \in 10, \ldots, 18$
– $T_0 = 8am$; $T = 8pm$

Concerning charging at the depot, prices for electricity are based on those provided by EDF (French Electricity Distribution company). Moreover, we generalized the E-VRPTW Benchmark Instances proposed in [14] and we tested our methods on those instances that include 100 customers and 21 charging stations. For all experiments, the parameter max_{Iter} was fixed at 10000 iterations, max_{impr} at 100 and $trial$ at 50. The LNS procedure uses first-improvement

---

Algorithm 3. ILS Algorithm

1: **Input:** A graph $G = (V', A)$ and a set of $m_{EV} + m_{CV}$ vehicles
2: **Output:** A set of routes assigned to at most $m_{EV} + m_{CV}$ vehicles
3: Let *Record* be the value of the best solution obtained
4: Initially, $Record = +\infty$, $impr = 0$, $restart = 0$, $n_{Iter} = 0$
5: **while** $restart < Max\_Restart$ **do**
6:     Generate an initial solution and let s_0 be this solution
7:     $s_1 \leftarrow IELS(s_0)$
8:     $record = cost_{s_1}$
9:     **while** $n_{Iter} < max_{Iter}$ AND $impr < max_{impr}$ **do**
10:         $s' \leftarrow Perturbation(s_1)$
11:         $s_1' \leftarrow IELS(s_1)$
12:         **if** $cost_{s_1'} < (1 + Dev) \times record$ **then**
13:             $s_1 \leftarrow s_1'$
14:         **end if**
15:         **if** $cost_{s_1'} < Record$ **then**
16:             $Record = obj(s_1')$, $impr = 0$
17:         **else**
18:             $impr + +$
19:         **end if**
20:         $n_{Iter} + +$
21:     **end while**
22:     $restart + +$
23: **end while**

---

algorithm. For each instance, we tested our methods following different implementation schemes obtained by varying the parameters of the ILS algorithm. The different implementation schemes are represented by the quintuplet (Num, $Num_{perturb}$, $Max\_Restart$, Dev, m) where $m = 0$ if the random insertion strategy is chosen and $m = 1$ if the regret insertion strategy is chosen. The computational results are summarized in Table 1. For each implementation scheme, the entries show the average gap (Gap(%)) and the average computational time (CPU (s)) for each instance category. The Gap of a generated solution (S) is calculated in relation to the initial solution ($Gap = \frac{S_{CRH} - S}{S}$). The computational results show that better solutions are obtained when the value $diff = Num_{Perturb} - Num$ is small ($diff = 1$, $diff = 2$) compared to the cases where $diff \geq 3$. Among the six first implementation schemes, the configuration (4,5,1,0,1) seems to be the best since it improve the initial solution by around 10 %. Moreover, we notice that the ILS with regret insertion strategy generates often better solutions than those generated by the ILS with random insertion strategy. Furthermore, the value of the deviation impacts the quality of the generated solutions. In fact, when the deviation is set to 0.1, the ILS algorithm improves by around 2 % the solution obtained in the case where the deviation is set to 0. However, in the case where $Dev = 0.2$, we notice a slight degradation of the obtained solutions. Finally, the ILS with restart improves the generated solutions by more than 80 % compared to the ILS without restart and the computational time remains acceptable. Thus, we can say that a good configuration to our algorithm is (4,5,2,0.1,1).

Table 1. Computational results of the Multi-Start ILS with Random and Regret Insertion Strategies.

Instance	(2,5,1,0,0)		(3,5,1,0,0)		(4,5,1,0,0)		(2,6,1,0,0)		(3,6,1,0,0)		(4,6,1,0,0)	
	Gap	CPU(s)	Gap	CPU(s)	Gap	CPU(s)	Gap	CPU(s)	Gap	CPU(s)	Gap	CPU(s)
Real	24.36	129.23	25.14	122.73	30.67	120.31	20.51	126.20	26.51	118.67	38.32	102.76
C1	23.18	12.85	24.66	16.97	32.30	13.50	21.35	14.08	25.33	11.80	32.45	11.69
C2	25.53	10.73	29.30	12.37	36.35	13.55	25.42	9.58	28.92	14.54	36.40	12.42
R1	35.43	13.98	40.29	12.14	45.23	12.77	32.41	14.34	40.29	13	44.53	13.22

Instance	(2,5,1,0,1)		(3,5,1,0,1)		(4,5,1,0,1)		(2,5,1,0.1,0)		(3,5,1,0.1,0)		(4,5,1,0.1,0)	
	Gap	CPU(s)	Gap	CPU(s)	Gap	CPU(s)	Gap	CPU(s)	Gap	CPU(s)	Gap	CPU(s)
Real	20.36	112.33	26.13	152.13	35.74	121.15	22.32	118.20	28.34	123.78	36.89	111.34
C1	20.90	12.39	25.50	10.45	34.52	13.99	22.08	14.59	27.37	9.59	35.05	11.59
C2	24.85	8.81	31.49	10.80	35.59	12.69	25.26	11.57	29.87	12.39	36.24	13.24
R1	32.89	16.12	41.16	11.02	45.04	13.74	34.82	14.01	42.81	11.02	47.99	13.24

Instance	(2,5,1,0.2,0)		(3,5,1,0.2,0)		(4,5,1,0.2,0)		(2,5,2,0,0)	
Real	22.36	112.43	25.08	131.42	31.74	123.35	82.92	195.20
C1	21.04	10.30	24.02	14.09	31.22	8.64	91.90	26.20
C2	25.91	11.31	29.82	14.32	35.78	12.36	102.35	21.02
R1	32.83	13.21	40.16	13.61	44.69	14.12	125.37	23.49

5   Conclusion

In this paper, we considered a new vehicle routing problem with mixed fleet of conventional and heterogenous electric vehicles and new real-life constraints. This problem consists in optimizing the routing of a set of vehicles with the objective of minimizing the overall routing and charging costs. To solve this problem, we developed a Multi-Start Iterated Local Search which uses a LNS with two insertion strategies in its Local Search procedure. All heuristic methods were tested on real data instances and generalized benchmark instances. As further work, we will test our methods on newly designed data instances and on other benchmark instances of some related problems. Next, we will relax our problem and compare our results with those of the literature. Moreover, we will consider other Local Search procedures and different perturbation mechanisms to improve our method.

References

1. Aissat, K., Oulamara, A.: A posteriori approach of real-time ridesharing problem with intermediate locations. In: proceedings of ICORES 2015 (2015)
2. Artmeier, A., Haselmayr, J., Leucker, M., Sachenbacher, M.: The optimal routing problem in the context of battery-powered electric vehicles. In: Workshop CROCS at CPAIOR-10, 2nd International Workshop on Constraint Reasoning and Optimization for Computational Sustainability (2010)

3. Bashash, S., Moura, S.J., Forman, J.C., Fathy, H.K.: Plug-in hybrid electric vehicle charge pattern optimization for energy cost and battery longevity. J. Power Sources **196**, 541–549 (2010)
4. Erdogan, S., Miller-Hooks, E.: A green vehicle routing problem. Transp. Res. Part E **48**, 100–114 (2012)
5. Felipe, Á., Ortuño, M.T., Righini, G., Tirado, G.: A heuristic approach for the green vehicle routing problem with multiple technologies and partial recharges. Transp. Res. Part E Logistics. Transp. Rev. **71**, 111–128 (2014)
6. Goeke, D., Schneider, M., Professorship, D.S.E.A.: Routing a mixed fleet of electric and conventional vehicles. Technical report, Darmstadt Technical University, Department of Business Administration, Economics and Law, Institute for Business Studies (BWL) (2014)
7. Hassin, R., Keinan, A.: Greedy heuristics with regret, with application to the cheapest insertion algorithm for the tsp. Oper. Res. Lett. **36**(2), 243–246 (2008)
8. Hiermann, G., Puchinger, J., Hartl, R.F.: The electric fleet size and mix vehicle routing problem with time windows and recharging stations. Technical report, Working Paper (2014). http://prolog.univie.ac.at/research/publications/downloads/Hie_2014_638.pdf. Accessed 17 July 2014
9. Lin, C., Choy, K., Ho, G., Chung, S., Lam, H.: Survey of green vehicle routing problem: Past and future trends. Expert. Syst. Appl. **41**, 1118–1138 (2014)
10. Lourenço, H.R., Martin, O.C., Stützle, T.: Iterated Local Search. Springer, New York (2003)
11. Millner, A.: Modeling lithium ion battery degradation in electric vehicles. In: 2010 IEEE Conference on Innovative Technologies for an Efficient and Reliable Electricity Supply (CITRES), pp. 349–356. IEEE (2010)
12. Pisinger, D., Ropke, S.: Large neighborhood search. Handbook of Metaheuristics, pp. 399–419. Springer, New York (2010)
13. Sassi, O., Oulamara, A.: Joint scheduling and optimal charging of electric vehicles problem. In: Murgante, B., Misra, S., Rocha, A.M.A.C., Torre, C., Rocha, J.G., Falcão, M.I., Taniar, D., Apduhan, B.O., Gervasi, O. (eds.) ICCSA 2014, Part II. LNCS, vol. 8580, pp. 76–91. Springer, Heidelberg (2014)
14. Schneider, M., Stenger, A., Goeke, D.: The electric vehicle routing problem with time windows and recharging stations. Technical report, University of Kaiserslautern, Germany (2012)
15. Shaw, P.: Using constraint programming and local search methods to solve vehicle routing problems. In: Maher, M.J., Puget, J.-F. (eds.) CP 1998. LNCS, vol. 1520, pp. 417–431. Springer, Heidelberg (1998)
16. Ramdane cherif, W., Haj Rachid, M., Bloch, C., Chatonnay, P.: New notation and classification scheme for vehicle routing problems. RAIRO (2014, to appear). doi:10.1051/ro/2014030

On the Complexity of Searching the Linear Ordering Problem Neighborhoods

Benjamin Correal$^{(\boxtimes)}$ and Philippe Galinier

École Polytechnique de Montréal, Montréal, Canada
benjamin.correal@gmail.com

Abstract. The linear ordering problem is an important and much studied NP-hard problem. The most efficient neighborhood for this problem is the so-called insert neighborhood. According to the literature, the best insert move can be found in $O(n^2)$. In this paper, we present a tree data structure that we name the maximum partial sum data structure. We show that using this data structure makes it possible to find, iteration after iteration, a best insert move in $O(n \log n)$ – after an initialization in $O(n^2)$. We also consider an alternative neighborhood named the interchange neighborhood. We show that this neighborhood can be searched in $O(n^2)$ – versus $O(n^3)$ in the best existing implementation.

1 Introduction

The linear ordering problem (LOP) can be defined as follows. Consider a set of n objects denoted by $1 \ldots n$ and assume that we must rank these objects. For any couple (i, j) of objects, such that $1 \leq i, j \leq n$ and $i \neq j$, there is a gain C_{ij} that is granted if object i comes before object j. The goal of the problem is to find a ranking (permutation) such that the sum of the gains is maximized – see a formal definition in Sect. 3.1. The LOP is an important NP-hard optimization problem that has applications in various fields, such as graph theory, economy, marketing, scheduling, and archeology [5].

Various exact algorithms have been proposed to solve the LOP, notably branch-and-bound [7], branch-and-cut [5] and a combined interior point/cutting plane algorithm [10]. Many different heuristics have also been proposed to tackle it, including pure neighborhood search techniques such as local search (hillclimbing) [2] and tabu search [8]; local search combined with perturbation mechanisms, such as iterated local search (ILS) [11,12] and variable neighborhood search (VNS) [4]; population-based heuristics hybridized with local search, such as path relinking [8], scatter search [1] and memetic algorithms (MA) [6,11,13]. According to a recent extensive experimental study [9], the most efficient heuristic is the memetic algorithm by Schiavinotto and Stützle, and the iterated local search proposed by the same authors comes a close second.

We can notice that the most efficient heuristics proposed to the LOP use neighborhood search as a major ingredient. It is the case of tabu search, but also of the heuristics that run repeatedly a hillclimbing operator, such as ILS, VNS or MA. Several neighborhoods have been proposed for the LOP, notably

© Springer International Publishing Switzerland 2015
G. Ochoa and F. Chicano (Eds.): EvoCOP 2015, LNCS 9026, pp. 150–159, 2015.
DOI: 10.1007/978-3-319-16468-7_13

the insert neighborhood and the interchange neighborhood. The insert neighborhood plays a central role because it is the one that is exploited by most efficient heuristics, if not all. Applying an insert move to a permutation consists in removing an element and inserting it at a different position. The size of this neighborhood is $(n-1)^2$. An alternative neighborhood is the so-called interchange neigborhood. An interchange move consists simply in swapping the position of two elements. The size of this neighborhood is $(n \times (n-1))/2$.

In this paper, we are interested by the complexity of searching the neighborhoods proposed to the LOP. This question is of course of great theoretical importance. In addition, it is critical in practice, as it may have a big impact on the speed of a neighborhood search heuristic. While early implementations searched the insert neighborhood in $O(n^3)$, Congram noticed that this task can be performed in $O(n^2)$ [3]. Directions about this implementation are given in [11]. In this paper, we present a technique that achieves this goal in only $O(n \log n)$. This result is obtained by using a tree structure that we name the maximum partial sum (MPS) data structure. Besides, it is indicated in the literature that the interchange neighborhood can be searched in $O(n^3)$ [11]. We show in the paper that this neighborhood can be searched in $O(n^2)$.

In the remaining of the paper, we first present our new MPS data structure (in Sect. 2). Then, we show how it can be used in order to implement the LOP insertion neighborhood (in Sect. 3). Additional results about the insert and the interchange neighborhoods follow (in Sect. 4). Finally, a conclusion and directions for future work are presented (in Sect. 5).

2   The Maximum Partial Sum Data Structure

In this section, we introduce an abstract data type that we name the maximum partial sum (MPS). Then, we describe an efficient implementation of the MPS.

2.1   Formal Definition

Let S be an ordered sequence of numbers: $(S_0, S_1, \ldots, S_{|S|-1})$. We will denote by $P_S : \{0 \ldots |S| - 1\} \to \mathbb{R}$ the function defined by

$$P_S(p) = \sum_{i=0}^{p} S_i \qquad (1)$$

This function will be named the *partial sum function* associated to S. We will use the following notations:

- $S.size = |S|$ denotes the cardinality of S.
- $S.sum = \sum_{i=0}^{|S|-1} S_i$ is the sum of the elements of S.
- $S.pmax = \max P_S() = \max\{P_S(p) : 0 \le p < |S|\}$.
- $S.argpmax = \operatorname{argmax} P_S()$ is an index value p ($0 \le p < |S|$) such that $P_S(p) = \max P_S()$.

We assume that S is first initialized, and that it thereafter undergoes a series of insertion and removal operations. Our goal is to determine the values of max $P_S()$ and argmax $P_S()$ after each new update operation. Therefore, the MPS must support the following operations.

- Init(L), for a list L: This operation initializes or reinitializes the value of S with the elements of L.
- Insert(val, j), for a number val and an index j $(0 \leq j < |S|)$: This operation inserts a new element val in position j.
- Remove(j), for an index j $(0 \leq j < |S|)$: This operation removes and returns the j-th element of S.
- GetMax(): This operation returns $S.pmax$ and $S.argpmax$.

In addition, we impose that the size of S will remain inferior to some integer n. In particular, the Init(.) operation creates an empty sequence if $|L| > n$ and the Insert(.) operation does not modify S if $|S| = n$.

It is clear that the following recurrence equations hold: $P_S(0) = S_0$ and $P_S(j) = P_S(j-1)+S_j$ for $j = 1 \ldots |S|-1$. Thanks to these equations, it is possible to compute the values of $S.pmax$ and $S.argpmax$ in $O(n)$ simply by computing step by step $P_S(j)$ for $j = 0 \ldots |S|-1$ and by updating the values of $S.pmax$ and $S.argpmax$ accordingly. In the next section, we will present an implementation that performs the initialization in $O(n)$ and each update operation in $O(\log n)$.

2.2 Implementation

Let us first present an important property that will be used in the following. Considering three sequences S, L and R such that $S = L \circ R$ (i.e., S is the concatenation of L and R) and $|L|, |R| \geq 1$, we claim that the following equations hold:

$$S.size = L.size + R.size \tag{2}$$

$$S.sum = L.sum + R.sum \tag{3}$$

$$S.pmax = \max(L.pmax, L.sum + R.pmax) \tag{4}$$

$$S.argpmax = \begin{cases} L.argpmax, & \text{if } S.pmax = L.pmax \\ L.size + R.argpmax, & \text{otherwise} \end{cases} \tag{5}$$

For example, let us consider the following sequences: $S = (5, -2, 1, 3, -4)$, $L = (5, -2)$ and $R = (1, 3, -4)$. We notice that $L.size = 2$, $L.sum = 3$, $L.pmax = 5$, $L.argpmax = 0$, and $R.size = 3$, $R.sum = 0$, $R.pmax = 4$ and $R.argpmax = 1$. As $S = L \circ R$, we can compute from the above equations that $S.size = 5$, $S.sum = 3$, $S.pmax = 7$ and $S.argpmax = 3$, which is actually correct.

The data structure we use is organized as a binary tree (a directed arborescence in which each internal node has exactly two sons) which corresponds to a recursive subdivision of S into sub-sequences. By abuse of language, we use the same notation for a node and for the sequence it represents. The root corresponds to S. A leaf corresponds to a subsequence of size 1, i.e. to a single element. Each internal node T has two sons (a left one denoted by $T.L$ and a right one denoted by $T.R$) such that $T = T.L \circ T.R$ and $|T.L|, |T.R| \geq 1$.

In each node T, we memorize the following information: $T.size$, $T.sum$, $T.pmax$, and $T.argpmax$. A node also has three links towards its father and its two sons. Note that the values contained in the sequence are not memorized in the nodes. An example of the tree is presented in Fig. 1.

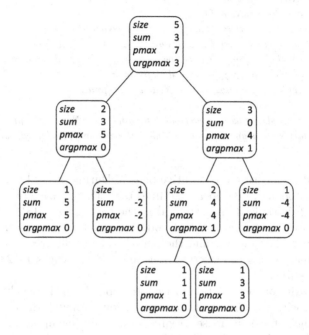

Fig. 1. Example of the data structure for a sequence $S = (5, -2, 1, 3, -4)$. The figure presents the values stored in each node.

The number of leaves in the tree equals $|S|$ and, as every internal node has two sons, the number of internal nodes equals $|S| - 1$. The size of the data structure is therefore $O(n)$.

The Init(.) operation builds a balanced tree that corresponds to the transmitted sequence. This can be done easily in $O(n)$ by recursively creating the nodes and initializing their fields according to Eqs. (2)–(5).

Let us now consider the Insert(.) procedure whose pseudocode is given below.

Procedure **Insert**(val, pos)
1. $currNode := root$
2.   **while** $currNode.size > 1$ **do**
3.   **if** $pos < currNode.left.size$ **then**
4. $currNode := currNode.left$
5.   **else**
6. $pos := pos - currNode.left.size$

7. $currNode := currNode.right$
8. $newNode := $ new node
9. $newNode.size := 1$
10. $newNode.sum := val$
11. $newNode.pmax := val$
12. $newNode.argpmax := 0$
13. $attach(newNode, currNode)$
14.   **while** $currNode \neq root$ **do**
15. $currNode := currNode.father$
16. $currNode.size := currNode.left.size + currNode.right.size$
17. $currNode.sum := currNode.left.sum + currNode.right.sum$
18.   **if** $currNode.left.pmax \geq currNode.left.sum + currNode.right.pmax$ **then**
19. $currNode.pmax := currNode.left.pmax$
20. $currNode.argpmax := currNode.left.argpmax$
21.   **else**
22. $currNode.pmax := currNode.left.sum + currNode.right.pmax$
23. $currNode.argpmax := currNode.left.size + currNode.right.argpmax$

The procedure has three consecutive phases. During phase 1, the position of the new node to be inserted is localized (lines 1–7). Then, during phase 2, the new node is created, initialized and attached in the tree (lines 8–13). The attach(.) procedure insures that every internal node has exactly two sons by creating a new node to join the new leaf and the one already in the tree. Finally, during phase 3, the ancestors of the inserted node are updated (lines 14–23) according to the Eqs. (2)–(5).

The complexity of the insertion procedure depends on the height of the tree. Unfortunately, with the described implementation, although the tree is initially balanced, its height can increase progressively following updating operations. Using more advanced well-known techniques makes it possible to maintain a balanced tree. In this case, the time complexity of the procedure will be $O(\log n)$. Finally, the removal of an element can be implemented similarly, with the same complexity. It is important to notice that, when the tree structure is updated, $S.pmax$ and $S.argpmax$ are accessible in constant time as these values are contained in the root node.

3   Implementing the Insertion Neighborhood

3.1   Definitions and Notation

A LOP problem instance is defined by an integer n and a n-by-n matrix C. A potential solution is any permutation π of $\{0, \ldots, n-1\}$. The score of a permutation is defined according to Eq. (6). The goal of the problem is to find a permutation of maximum score.

$$f(\pi) = \sum_{i=0}^{n-2} \sum_{j=i+1}^{n-1} C_{\pi_i \pi_j} \tag{6}$$

We consider two types of moves: insert moves and interchange moves. Considering a reference configuration π and a move m (whether it is an insert move or an interchange move), we will denote by $\pi \oplus m$ the configuration obtained by applying move m to π.

An insert move will be denoted by $(i \to j)$, where $0 \le i, j < n$, and $i \ne j$. Applying move $(i \to j)$ to configuration π consists in moving element π_i from its current position to a new position j, as expressed by Eq. (7).

$$\pi \oplus (i \to j) = \begin{cases} (\pi_0, \dots, \pi_{i-1}, \pi_{i+1}, \dots, \pi_j, \pi_i, \pi_{j+1}, \dots, \pi_n), & i < j \\ (\pi_0, \dots, \pi_{j-1}, \pi_i, \pi_j, \dots, \pi_{i-1}, \pi_{i+1}, \dots, \pi_n), & i > j \end{cases} \tag{7}$$

An interchange move will be denoted by $(i \leftrightarrow j)$, where $0 \le i, j < n$, and $i \ne j$. Applying move $(i \leftrightarrow j)$ to configuration π consists in exchanging the positions of elements π_i and π_j, as expressed by Eq. (8) – assuming, without loss of generality, that $i < j$.

$$\pi \oplus (i \leftrightarrow j) = (\pi_0, \dots, \pi_{i-1}, \pi_j, \pi_{i+1}, \dots, \pi_{j-1}, \pi_i, \pi_{j+1}, \dots, \pi_n) \tag{8}$$

Given a move m, we denote by $\delta(m)$ the performance of the move, i.e. its impact on the score function, as expressed by Eq. (9).

$$\delta_\pi(m) = f(\pi \oplus m) - f(\pi) \tag{9}$$

In order to simplify the notation used in the next procedures and definitions, we use the following notation – see Eq. (10). Notice that $\Delta(\pi, i, i+1)$ is the score of the move $(i \to i+1)$.

$$\Delta(\pi, i, j) = C_{\pi_j \pi_i} - C_{\pi_i \pi_j} \tag{10}$$

3.2   The Regular $O(n^2)$ Implementation

We name findBestMove(.) the procedure that returns the best insert move. The pseudocode of this procedure is given below. This implementation is consistent with the description given in [11]. The findBestMove(.) procedure simply calls repeatedly the findBestNewPosition(.) procedure; the role of this latter procedure is to return the best new position j for the element whose index i is transmitted in parameter, along with the score of the corresponding move $(i \to j)$.

Procedure **findBestMove**()
1. $deltaMax := -\infty$
2.   **for** $i := 0 \dots n - 1$ **do**
3. $(delta, j) :=$ **findBestNewPosition**(i)
4.   **if** $delta > deltaMax$ **then**
5. $deltaMax := delta$

6. $ibest := i$
7. $jbest := j$
8.   **return** $(delta, ibest, jbest)$

A discussion and hints about how to implement the second procedure are given in [11]. We believe it may be helpful for the reader to give explicitly the pseudocode of the implementation.

Procedure **findBestNewPosition**(i)
1. $deltaMax := -\infty$
2. $delta := 0$
3.   **for** $j := i - 1 \ldots 0$ **do**
4. $delta := delta + \Delta(\pi, j, i)$
5.   **if** $delta > deltaMax$ **then**
6. $deltaMax := delta$
7. $jbest := j$
8. $delta := 0$
9.   **for** $j := i + 1 \ldots n - 1$ **do**
10. $delta := delta + \Delta(\pi, i, j)$
11.   **if** $delta > deltaMax$ **then**
12. $deltaMax := delta$
13. $jbest := j$
14.   **return** $(delta, jbest)$

The procedure consists mainly in two successive phases, each phase corresponding to a loop. During the first phase (lines 1–7), the score of each move $(i \to j)$ is computed for decreasing values of j, from $j = i - 1$ to $j = 0$. Then, during the second phases (lines 8–13), it is computed for increasing values of j, from $j = i + 1$ to $j = n - 1$. It is clear that the findBestNewPosition(.) is O(n). As a result, computing the best move can be performed in $O(n^2)$ by using the findBestMove(.) procedure – as already stated in [11].

3.3   Finding the Best Insert Move in $O(n \log n)$

From the above pseudocode, let us observe what happens when findBestNew Position(i) is called. We can notice that each loop computes in fact the *pmax* and the *argpmax* values of a particular sequence of numbers in each of the two phases; we denote these sequences by Neg_i and Pos_i:

$$Neg_i = \big(\Delta(\pi, i - 1, i), \Delta(\pi, i - 2, i), \ldots, \Delta(\pi, 0, i)\big)$$
$$Pos_i = \big(\Delta(\pi, i, i + 1), \Delta(\pi, i, i + 2), \ldots, \Delta(\pi, i, n - 1)\big)$$

In our implementation, we will use the MPS tree data structure presented above. For each $i = 0 \ldots n - 1$, we use two MPS data structures denoted by DN_i and DP_i associated to Neg_i and Pos_i. These data structures are initialized according to the initial configuration π and then updated throughout the search.

After a move $(i_m \rightarrow j_m)$ has been performed, the data structures are updated according to the updateAfterMove(i_m, j_m) procedure whose pseudocode is given below. The procedure findBestNewPosition(i) can then be replaced with a $O(1)$ choice between the best move of both sequences ($DN_i.GetMax()$ and $DP_i.GetMax()$), reducing the complexity of findBestMove(.) to $O(n)$.

Procedure **updateAfterMove**(i_m, j_m)
1.   **for** $i := 0 \ldots n - 1$, **if** $i \neq i_m$ **do**
2.   **if** $i_m < i$ **then**
3. $v = DN_i.\text{Remove}(i_m)$
4.   **else**
5. $v = DP_i.\text{Remove}(i_m - i - 1)$
6.   **if** $j_m < i$ **then**
7. $DN_i.\text{Insert}(v, j_m)$
8.   **else**
9. $DP_i.\text{Insert}(v, j_m - i - 1)$
10. $DN_{i_m}.\text{Init}(Neg_{i_m})$
11. $DP_{i_m}.\text{Init}(Pos_{i_m})$

First, the procedure updates DN_i and DP_i, for every value of i different from i_m. Then, DN_{i_m} and DP_{i_m} are rebuilt from scratch using the updated values of Neg_{i_m} and Pos_{i_m}. As each updating procedure is performed in $O(\log n)$ and the initialization procedure in $O(n)$, the overall complexity of the procedure is $O(n \log n)$. This is also the complexity of a whole local search iteration when using the proposed implementation.

In summary, the size complexity of the new implementation is $O(n^2)$, which is the same as the size complexity of the input matrix. Using the new implementation, the complexity of the initialization is $O(n^2)$; then finding and applying the best move on each local search iteration costs $O(n \log n)$, versus $O(n^2)$ when using the implementation presented in the literature.

4   Additional Results About the Two Neighborhoods

It is easy to remark that the score of an interchange move can be computed by summing the scores of two insert moves according to the following equation:

$$\forall i, j \ s.t. \ i < j, \delta_\pi(i \leftrightarrow j) = \delta_\pi(i \rightarrow j) + \delta_\pi(j \rightarrow i + 1) \tag{11}$$

First note that, in Eq. (11), i and j do not play the same role, since we assume that $i < j$. The property expressed by this equation has two important consequences. The first consequence is related to the complexity of a procedure for searching the interchange neighborhood. The second consequence is related to the set of local optima of the two neighborhoods (insert and interchange).

4.1   Finding the Best Interchange Move in $O(n^2)$

It is possible to explore efficiently the interchange neighborhood by using the following technique. For this implementation, it is required to use a n-by-n matrix. First store in the matrix the score of each insert move. Then compute in constant time the score of each interchange move according to Eq. (11). It is clear that this technique is $O(n^2)$, versus $O(n^3)$ as it is reported in the literature [11].

4.2   About the Local Optima of the Two Neigborhoods

Let us denote by $LO_\rightarrow$ and $LO_\leftrightarrow$ the sets of local optima with respect to the insert and the interchange neighborhoods, respectively. The property expressed by Eq. (12) holds.

$$LO_\rightarrow \subseteq LO_\leftrightarrow \tag{12}$$

The proof of this property is as follows. Let us assume that $\pi \in LO_\rightarrow$. Therefore, we have that $\delta_\pi(i \rightarrow j)$ is non-positive, for every couple (i, j). As a result, for every i and j, we must have that $\delta_\pi(i \leftrightarrow j)$ is non-positive, as it is the sum of two non-positive numbers, according to Eq. (11). Thus, $\pi \in LO_\leftrightarrow$.

5   Discussion and Conclusion

In this paper, we have presented results related to the two main neighborhoods proposed to the Linear Ordering Problem, namely the insert and the swap neighborhoods. The three main contributions of our paper are the following:

1. Finding the best insert move on each iteration can be done in $O(n \log n)$.
2. Finding the best swap move can be done in $O(n^2)$.
3. The set of local optima w.r.t. the insert neighborhood is included in the set of local optimal w.r.t. the swap neighborhood.

Our most interesting contribution may be the improvement of the complexity of searching the insert neighborhood. The new (worst-case) complexity we obtain ($O(n \log n)$) represents a significant improvement over the best complexity known so far ($O(n^2)$). The technique we propose relies on a tree data structure that we name the maximum partial sum structure. This result is of course important for theoretical reasons. It may also be very useful in practice. It could be used as is, in order to develop a hill-climbing heuristic that performs the best improvement policy. More interestingly, it could also be easily adapted in order to implement a tabu algorithm. Developing a tabu algorithm dotted with the maximum partial sum structure will be the focus of our future work.

Our third contribution is related to the sets of local optima of the two neigbborhoods. Empirically, the insert neighborhood has been observed to be more efficient than the swap neighborhood. In fact, the insert neighborhood is the one used in the best-performing LOP heuristics of the literature. This superiority is also confirmed by systematic experiments performed in [9]. Our result provides for the first time a clear theoretical basis that explains this superiority.

References

1. Campos, V., Laguna, M., Martí, R.: An experimental evaluation of a scatter search for the linear ordering problem. J. Global Optim. **21**, 397–414 (2001)
2. Chanas, S., Kobylànski, P.: A new heuristic algorithm solving the linear ordering problem. Comput. Optim. Appl. **6**, 191–205 (1996)
3. Congram, R.K.: Polynomially searchable exponential neighbourhoods for sequencing problems in combinatorial optimisation. Ph.D. thesis, University of Southampton, Faculty of Mathematical Studies, UK (2000)
4. García, C.G., Pérez-Brito, D., Campos, V., Martí, R.: Variable neighborhood search for the linear ordering problem. Comput. Oper. Res. **33**, 3549–3565 (2006)
5. Grötschel, M., Jünger, M., Reinelt, G.: A cutting plane algorithm for the linear ordering problem. Oper. Res. **32**(6), 1195–1220 (1984)
6. Huang, G., Lim, A.: Designing a hybrid genetic algorithm for the linear ordering problem. In: Cantú-Paz, E., et al. (eds.) GECCO 2003. LNCS, vol. 2723, pp. 1053–1064. Springer, Heidelberg (2003)
7. Kaas, R.: A branch and bound algorithm for the acyclic subgraph problem. Eur. J. Oper. Res. **8**, 355–362 (1981)
8. Laguna, M., Martí, R., Campos, V.: Intensification and diversification with elite tabu search solutions for the linear ordering problem. Comput. OR **26**(12), 1217–1230 (1999)
9. Martí, R., Reinelt, G., Duarte, A.: A benchmark library and a comparison of heuristic methods for the linear ordering problem. Comput. Optim. Appl. **51**(3), 1297–1317 (2012)
10. Mitchell, J.E., Borchers, B.: Solving Linear Ordering Problems with a Combined Interior Point/Simplex Cutting Plane Algorithm. Kluwer Academic Publishers, Dordrecht (2000)
11. Schiavinotto, T., Stützle, T.: The linear ordering problem: instances, search space analysis and algorithms. J. Math. Model. Algorithms **3**(4), 367–402 (2004)
12. Valdez, G.C., Bastiani-Medina, S.S.: Iterated local search for the linear ordering problem. Int. J. Comb. Optim. Probl. Inf. **3**(1), 12–20 (2012)
13. Ye, T., Wang, T., Lü, Z., Hao, J.K.: A Multi-parent Memetic Algorithm for the Linear Ordering Problem. CoRR abs/1405.4507 (2014)

Runtime Analysis of $(1 + 1)$ Evolutionary Algorithm Controlled with Q-learning Using Greedy Exploration Strategy on ONEMAX+ZEROMAX Problem

Denis Antipov[1], Maxim Buzdalov[1]([✉]), and Benjamin Doerr[2]

[1] ITMO University, 49 Kronverkskiy av., Saint-Petersburg, Russia, 197101
antipovden@yandex.ru, mbuzdalov@gmail.com
[2] LIX, École Polytechnique, 91128 Palaiseau Cedex, France
doerr@lix.polytechnique.fr

Abstract. There exist optimization problems with the target objective, which is to be optimized, and several extra objectives. The extra objectives may or may not be helpful in optimization process in terms of the number of objective evaluations necessary to reach an optimum of the target objective.

ONEMAX+ZEROMAX is a previously proposed benchmark optimization problem where the target objective is ONEMAX and a single extra objective is ZEROMAX, which is equal to the number of zero bits in the bit vector. This is an example of a problem where extra objectives are not good, and objective selection methods should ignore the extra objectives. The EA+RL method is a method which selects objectives to be optimized by evolutionary algorithms (EA) using reinforcement learning (RL). Previously it was shown that it runs in $\Theta(N \log N)$ on ONEMAX+ZEROMAX when configured to use the randomized local search algorithm and the Q-learning algorithm with the greedy exploration strategy.

We present the runtime analysis for the case when the $(1 + 1)$-EA algorithm is used. It is shown that the expected running time is at most $3.12eN \log N$.

1 Introduction

Single-objective optimization can often benefit from multiple objectives [6,7,9]. Different approaches are known from the literature. Some researchers introduce additional objectives to escape from the plateaus [1]. Decomposition of the primary objective into several objectives also helps in many problems [5,7]. Additional objectives may also arise from the problem structure [8].

1.1 Approaches for Extra Objectives

Different approaches may be applied to a problem with the "original" objective, which can be called the *target* objective, and some extra objectives.

G. Ochoa and F. Chicano (Eds.): EvoCOP 2015, LNCS 9026, pp. 160–172, 2015.
DOI: 10.1007/978-3-319-16468-7_14

The *multi-objectivization* approach is to optimize *all* extra objectives at once using a multi-objective optimization algorithm [5, 7]. The *helper-objective* approach is to optimize simultaneously the target objective and some (not necessarily all, in some cases, only one is preferrable) extra objectives, switching between them from time to time [6].

The approaches above are designed in the assumption that the extra objectives are crafted to help optimizing the target objective. However, this is not always true, especially when their properties are unknown. In fact, the extra objectives may support or obstruct the process of optimizing the target objective. The EA+RL method [3] was developed to cope with such situations. The idea of this method is to use a single-objective optimization algorithm and switch between the objectives (which include the target one and the extra ones). To find the most suitable objective for the optimization, reinforcement learning algorithms are used [10].

1.2   Reinforcement Learning and EA+RL

Reinforcement learning [10] uses the concepts of *state*, *action* and *reward*. A reinforcement learning algorithm is often thought to control an *agent* which interacts with a certain *environment*. The agent receives the current state from the environment as input. It should return an action to apply on the environment. For that action, it receives a reward. The aim of the reinforcement learning algorithm is to maximize the total reward by choosing appropriate actions in different states. The total reward can be treated as a sum of all rewards received by the algorithm, or as a *discounted* reward, when a reward for i-th step from the end is taken with a weight of γ^i, where $0 < \gamma < 1$.

In the EA+RL method, actions are objectives to choose, while states and rewards are defined depending on the problem. A good choice of the reward can be the value of the target objective after the selection of an objective minus the value of the target objective before it. The sum of all rewards during the optimization is equal to the difference between the final value of the target objective and its initial value, so optimization of the reward leads to optimization of the target objective.

2   Analyzed Problem and Algorithm

ONEMAX is a well-known optimization problem widely used in theoretical research on evolutionary computation. It can be defined as "maximize the number of one-bits in a bit vector of length N". It is known that simple evolutionary algorithms, such as randomized local search (RLS) or $(1 + 1)$ evolutionary algorithm $((1 + 1)$-EA), solve this problem in $\Theta(N \log N)$ function evaluations [11].

We define ZEROMAX, a counterpart of ONEMAX, as follows: the number of zero-bits in a bit vector needs to be maximized. Clearly, the maximum point of ZEROMAX is the same as the minimum point of ONEMAX, and vice versa. Moreover, any change that increases the ONEMAX fitness decreased the ZEROMAX fitness at the same time.

In paper [2], ONEMAX+ZEROMAX was defined as an optimization problem with extra objectives where ONEMAX is the *target* objective and ZEROMAX is the extra objective. Clearly, for this problem every objective selection algorithm should eventually manage to ignore the offensive extra objective. It was shown that the EA+RL method [3], the objective selection method based on reinforcement learning, indeed learns to ignore the ZEROMAX objective when randomized local search (RLS) is used as an optimizer and finds the ONEMAX optimum in $\Theta(N \log N)$, more precisely, at most twice slower than RLS itself does when optimizing ONEMAX. The proof in [2] was done by analysing the Markov chain which modeled the optimization process. Since this Markov chain possessed a simple linear structure, the proof was relatively easy. However, when the optimizer is changed to $(1 + 1)$-evolutionary algorithm, which is able to flip more than one bit at a time, using Markov chains becomes insanely complicated.

In this paper, we consider the $(1 + 1)$ evolutionary algorithm with the fixed probability of flipping a bit $p = 1/N$ as a single-objective optimization algorithm. To select which objective to optimize at each iteration, we use the EA+RL method [3], which internally uses a reinforcement learning (RL) algorithm [10] to do this. The *actions* of the RL algorithm are the possible objectives, so the set of possible actions for the considered problem is $\{$ONEMAX, ZEROMAX$\}$. The choices for the EA+RL method are fixed in this paper to the following values:

- RL algorithm: the Q-learning algorithm with greedy exploration strategy [10];
  - the learning rate: an arbitraty $\alpha \in (0; 1)$;
  - the discount factor: an arbitrary $\gamma \in [0; 1]$;
- RL state: the value of the ONEMAX fitness;
- RL reward for the action: the difference of the ONEMAX fitness values after the action and before the action.

The pseudocode for the $(1 + 1)$-EA controlled by the EA+RL method with the parameters listed above is shown in Fig. 1.

3 Learning Lemma

The Q-learning algorithm stores estimations of action rewards as a $Q(s, a)$ matrix, where $Q(s, a)$ is the expected reward for applying an action a in a RL state s. When randomized local search is used, it was shown in [2] that EA+RL learns to ignore the offensive ZEROMAX objective in the following way: once it leaves each RL state for the first time, it maintains $Q(s, 1) > Q(s, 0)$, which makes it select ONEMAX each time it enters the same state the next time. The same idea is true in the current situation, but in a more complicated manner.

Lemma 1 (Learning lemma). *If the algorithm has not reached the terminal state (where $s = N$) and $\gamma < \frac{1}{N-1}$ then for every non-terminal state s it is true that:*

$$Q(s, 0) \leq 0 \leq Q(s, 1) < N - 1 - s.$$

Additionally, if the algorithm has never left a state s, it is true that $Q(s, 0) = Q(s, 1) = 0$. Otherwise, $Q(s, 0) < Q(s, 1)$.

```
1:  X ← current individual, vector of N zeros
2:  Q ← transition quality matrix, N × 2, filled with zeros
3:  f₁ ← ONEMAX, the target objective
4:  f₀ ← ZEROMAX, the extra objective
5:  MUTATE(X) ← mutation operator: inverts each bit with p = 1/N
6:  while f₁(X) < N do
7:      s ← f₁(X)
8:      Y ← MUTATE(X)
9:      f, i: chosen fitness function and its index
10:     if Q(s, 0) > Q(s, 1) then
11:         i ← 0
12:     else if Q(s, 0) < Q(s, 1) then
13:         i ← 1
14:     else
15:         i ← RANDOM(0, 1)
16:     end if
17:     f ← fᵢ
18:     if f(Y) ≥ f(X) then
19:         X ← Y
20:     end if
21:     s' ← f₁(X)
22:     r ← s' − s
23:     Q(s, i) ← (1 − α)Q(s, i) + α(r + γ · maxⱼ Q(s', j))
24: end while
```

Fig. 1. $(1 + 1)$-EA controlled by EA+RL using the greedy Q-learning algorithm

Proof. We use mathematical induction. The base is obvious: in the very beginning, all $Q(i, j)$ are zeros and the algorithm has never left any state, so the lemma statement is true. Assume that the lemma statement was true for all previous algorithm iterations. The current iteration can have the following forms:

1. The algorithm has never left the current state and remains there.
2. The algorithm has left the current state for the first time.
3. The algorithm has left the current state before.

Below we denote the state before the current iteration as s, the state after the current iteration as s', the Q-values before the current iteration as $Q(i, j)$ and the Q-values after the current iteration as $Q'(i, j)$.

Case 1. By induction hypothesis, $Q(s, 0) = Q(s, 1) = 0$. If the algorithm remains at the state s, the reward is zero, so all the components of the expression at line 23 are zeros. Whatever objective i was selected, $Q'(s, i)$ is set to zero, and $Q'(s, 1 - i)$ remains zero as well, so the induction hypothesis is proven for the current iteration.

Case 2. By induction hypothesis, $Q(s, 0) = Q(s, 1) = 0$. This means that the objective is ONEMAX with the probability of 0.5 and ZEROMAX with the probability of 0.5. As $s' \neq s$, the change was accepted by the selected objective, so for ONEMAX ($i = 1$) $s' \geq s + 1$, while for ZEROMAX ($i = 0$) $s' \leq s - 1$.

By induction hypothesis, $Q(s', 0) \leq 0 \leq Q(s', 1) < N - s' - 1$. This means that for the chosen objective i the following will be true:

$$Q'(s, i) = (1 - \alpha)Q(s, i) + \alpha \left(s' - s + \gamma \max_j Q(s', j) \right) = \alpha(s' - s + \gamma Q(s', 1)).$$

The upper bound on $s' - s + \gamma Q(s', 1)$, provided that $s' < N$, is:

$$s' - s + \gamma Q(s', 1) < s' - s + \frac{N - s' - 1}{N - 1} = s' \left(1 - \frac{1}{N - 1} \right) + 1 - s$$

$$= s' \frac{N - 2}{N - 1} + 1 - s \leq \frac{(N - 1)(N - 2)}{N - 1} + 1 - s = N - s - 1.$$

It follows that $Q'(s, i) < N - s - 1$ as well. The lower bound is $Q'(s, i) \geq \alpha(s' - s)$. For $i = 1$ these two bounds immediately yield that $0 < Q'(s, 1) < N - s - 1$. For $i = 0$ we should additionally use the fact that $s' \leq s - 1$, which brings:

$$Q'(s, 0) < s' \frac{N - 2}{N - 1} + 1 - s \leq s' + 1 - s \leq 0.$$

To sum up, after an iteration which leaves a state s for the first time it will be that either $0 < Q'(s, 1) < N - s - 1$ and $Q'(s, 0) = 0$ or $Q'(s, 0) < 0$ and $Q'(s, 1) = 0$. This proves the induction hypothesis for the current iteration in the considered case.

Case 3. By induction hypothesis, $Q(s, 0) < Q(s, 1)$, so the ONEMAX objective is selected. As a result, $s' \geq s$. Using the upper bound on $s' - s + \gamma Q(s', 1)$ proven in the previous case (which still holds under assumptions of the current case), the fact that it is non-negative and the induction assumption that $Q(s', 0) \leq Q(s', 1)$, we get the bounds on $Q'(s, 1)$:

$$Q'(s, 1) = (1 - \alpha)Q(s, 1) + \alpha \left(s' - s + \gamma Q(s', 1) \right)$$
$$< (1 - \alpha)(N - s - 1) + \alpha(N - s - 1) = N - s - 1.$$
$$Q'(s, 1) \geq (1 - \alpha)Q(s, 1).$$

In any case, $Q'(s, 1) < N - s - 1$. If $Q(s, 1) > 0$, then $Q'(s, 1) > 0$. If $Q(s, 1) = 0$, then $Q'(s, 1) \geq 0$. This proves the induction hypothesis for the current iteration in the considered case.

In all three possible cases the induction hypothesis is proven, which completes the proof.   □

This lemma lets us describe each RL state in any moment of time either as *learned* or *unlearned*. In the learned state the algorithm always selects the correct objective, ONEMAX. In the unlearned state, it selects either ONEMAX or ZEROMAX with equal probabilities. Each unlearned state becomes learned when the algorithm leaves this state and enters another one. All these considerations are true when $\gamma < 1/(N - 1)$, so we consider it to be so in the rest of the paper until explicitly noted.

4 Transition Probabilities

What is the exact probability that the independent bit-flip mutation (with a probability of flipping each bit equal to $1/N$) constructs a bit string with j one-bits from a bit string with i one-bits? Consider the situation where $i < j$: this means that $j - i + k$ zeros and k ones are flipped. The exact expressions for these probabilities are given below.

$$
P^{i,j} = \begin{cases}
\sum\limits_{k=0}^{\min(N-j,i)} \binom{N-i}{j-i+k}\binom{i}{k} \left(\frac{1}{N}\right)^{j-i+2k} \left(1 - \frac{1}{N}\right)^{N-(j-i+2k)} & \text{if } i < j, \\[2ex]
\sum\limits_{k=0}^{\min(N-i,j)} \binom{i}{i-j+k}\binom{N-i}{k} \left(\frac{1}{N}\right)^{i-j+2k} \left(1 - \frac{1}{N}\right)^{N-(i-j+2k)} & \text{if } i > j, \quad (1) \\[2ex]
\sum\limits_{k=0}^{\min(N-i,i)} \binom{N-i}{k}\binom{i}{k} \left(\frac{1}{N}\right)^{2k} \left(1 - \frac{1}{N}\right)^{N-2k} & \text{if } i = j.
\end{cases}
$$

In an unlearned state, ONEMAX or ZEROMAX is chosen with the probability of 0.5. Together with the probabilities given above, transition probability $P_U^{i,j}$ from an unlearned state i to a state j is $\frac{1}{2}P^{i,j}$ if $i \neq j$ and $1 - \frac{1}{2}\sum_{k \neq i} P^{i,k} = \frac{1}{2}\left(1 + P^{i,i}\right)$ if $i = j$.

In a learned state, ONEMAX is always chosen, so the transition probability $P_L^{i,j}$ from a learned state i to a state j is $P^{i,j}$ if $i < j$, $1 - \sum_{k=i+1}^{j} P^{i,j}$ if $i = j$ and zero otherwise.

4.1 Lower and Upper Bound on $P^{i,j}$

The expressions for $P^{i,j}$ are rather complex. The following theorem gives a lower and an upper bound on $P^{i,j}$.

Theorem 1. *Assume that $i \neq j$. Let $S^{i,j}$ be the following:*

$$
S^{i,j} = \begin{cases}
\binom{N-i}{j-i} \left(\frac{1}{N}\right)^{j-i} \left(1 - \frac{1}{N}\right)^{N-(j-i)} & \text{if } i < j, \\[2ex]
\binom{i}{i-j} \left(\frac{1}{N}\right)^{i-j} \left(1 - \frac{1}{N}\right)^{N-(i-j)} & \text{if } i > j.
\end{cases} \quad (2)
$$

Then $S^{i,j} \leq P^{i,j} \leq \frac{8}{7} S^{i,j}$.

Proof. The lower bounds are proven easily, since $S^{i,j}$ are the addends for $k = 0$ in (1), and all these addends are positive.

We denote as $S_k^{i,j}$ the k-th addend of the sum in (1) corresponding to $P^{i,j}$. Specifically, $S^{i,j} = S_0^{i,j}$. Consider the case of $i < j$. The ratio of the k-th addend to the $(k+1)$-th addend is:

$$
\frac{S_k^{i,j}}{S_{k+1}^{i,j}} = \frac{\binom{N-i}{j-i+k}\binom{i}{k} \left(\frac{1}{N}\right)^{j-i+2k} \left(1 - \frac{1}{N}\right)^{N-(j-i+2k)}}{\binom{N-i}{j-i+k+1}\binom{i}{k+1} \left(\frac{1}{N}\right)^{j-i+2k+2} \left(1 - \frac{1}{N}\right)^{N-(j-i+2k)-2}}
$$

$$
= \frac{(j-i+k+1)(k+1)}{(N-j-k)(i-k)} N^2 \left(1 - \frac{1}{N}\right)^2 = \frac{(j-i+k+1)(k+1)}{(N-j-k)(i-k)} (N-1)^2.
$$

When i and j are fixed, this ratio grows as k grows, so

$$\frac{S_k^{i,j}}{S_{k+1}^{i,j}} \geq \frac{S_0^{i,j}}{S_1^{i,j}} = \frac{j-i+1}{(N-j)i}(N-1)^2.$$

When i is fixed, this ratio grows as j grows, so we replace j with its minimum possible value $i+1$ and then minimize the result with $i = \frac{N-1}{2}$:

$$\frac{S_k^{i,j}}{S_{k+1}^{i,j}} > \frac{2(N-1)^2}{(N-i-1)i} \geq \frac{2(N-1)^2}{\frac{N-1}{2}\frac{N-1}{2}} = \frac{8(N-1)^2}{(N-1)^2} = 8.$$

This means that $P^{i,j}$ can be bounded by a sum of geometric progression:

$$P^{i,j} = \sum_{k=0}^{\min(N-j,i)} S_k^{i,j} \leq \sum_{k=0}^{\min(N-j,i)} \left(\frac{1}{8}\right)^k S_0^{i,j} \leq \sum_{k=0}^{\infty} \left(\frac{1}{8}\right)^k S_0^{i,j} = \frac{8}{7}S_0^{i,j} = \frac{8}{7}S^{i,j}.$$

The case of $i > j$ is proven in the similar way. □

In the rest of the paper, we denote the $\frac{8}{7}$ constant from this theorem by R.

4.2   Lower and Upper Bounds on Partial Sums of $P^{i,j}$

Theorem 2. *If* $V_i = \left(\frac{N-1}{N}\right)^{N-i} - \left(\frac{N-1}{N}\right)^N$, *then* $V_i \leq \sum_{j=0}^{i-1} P^{i,j} \leq RV_i$.

Proof. Considering the definition of $S^{i,j}$ from Theorem 1, we get that:

$$\sum_{j=0}^{i-1} S^{i,j} = \sum_{j=0}^{i-1} \binom{i}{i-j} \left(\frac{1}{N}\right)^{i-j} \left(1-\frac{1}{N}\right)^{N-(i-j)}$$

$$= \left(1-\frac{1}{N}\right)^N \sum_{j=0}^{i-1} \binom{i}{i-j} \left(\frac{1}{N-1}\right)^{i-j}$$

$$= \left(1-\frac{1}{N}\right)^N \left(\left(\frac{N}{N-1}\right)^i - 1\right) = V_i.$$

As $P^{i,j} \geq S^{i,j}$, it follows that $\sum_{j=0}^{i-1} P^{i,j} \geq \sum_{j=0}^{i-1} S^{i,j} = V_i$. Similarly, as $P^{i,j} \leq RS^{i,j}$, it follows that $\sum_{j=0}^{i-1} P^{i,j} \leq R \sum_{j=0}^{i-1} S^{i,j} = RV_i$. □

Theorem 3. *If* $W_i = \left(\frac{N-1}{N}\right)^i - \left(\frac{N-1}{N}\right)^N$, *then* $W_i \leq \sum_{j=i+1}^{N} P^{i,j} \leq RW_i$.

4.3   Lower and Upper Bounds on Other Expressions

Theorem 4. *If* $Y_i = \frac{i}{N-1}\left(\frac{N-1}{N}\right)^{N-i+1}$, *then* $Y_i \leq \sum_{j=0}^{i-1} P^{i,j}(i-j) \leq RY_i$.

Proof. Consider $\sum_{j=0}^{i-1} S^{i,j}(i-j)$:

$$
\sum_{j=0}^{i-1} S^{i,j}(i-j) = \sum_{j=0}^{i-1} \binom{i}{i-j} \left(\frac{1}{N}\right)^{i-j} \left(1 - \frac{1}{N}\right)^{N-(i-j)} (i-j)
$$

$$
= \left(1 - \frac{1}{N}\right)^N \sum_{j=0}^{i-1} \binom{i}{j} \frac{i-j}{(N-1)^{i-j}}
$$

$$
= \left(1 - \frac{1}{N}\right)^N (1-N) \sum_{j=0}^{i-1} \binom{i}{j} \left(\frac{1}{(N-1)^{i-j}}\right)'_N
$$

$$
= \left(1 - \frac{1}{N}\right)^N (1-N) \left(\sum_{j=0}^{i-1} \binom{i}{j} \frac{1}{(N-1)^{i-j}}\right)'_N
$$

$$
= \left(1 - \frac{1}{N}\right)^N (1-N) \left(\left(\frac{N}{N-1}\right)^i - 1\right)'_N
$$

$$
= \left(1 - \frac{1}{N}\right)^N \frac{i}{N-1} \left(\frac{N}{N-1}\right)^{i-1} = \frac{i}{N-1} \left(\frac{N-1}{N}\right)^{N-i+1} = Y_i.
$$

Similarly to Theorem 2, we prove the bounds on the required sum.   □

Theorem 5. *If* $Z_i = \frac{N-i}{N-1} \left(\frac{N-1}{N}\right)^{i+1}$, *then* $Z_i \leq \sum_{j=i+1}^N P^{i,j}(j-i) \leq RZ_i$.

5   Drift Analysis

We analyse the running time of the algorithm using the additive drift theorem [4]. To do that, we construct the following potential function:

$$
\Phi(i,l) = \sum_{t=i}^{N-1} \frac{N}{N-t} + CN \sum_{t=0}^{N-1} \frac{1-l(t)}{N-t},
$$

where i is the current state (equal to the number of one-bits), $l(t)$, the *learn indicator*, is equal to one if the state t is a learned state and to zero otherwise, and C is a constant.

Such function rewards the algorithm not only for getting closer to the optimum, but for learning a state as well. Note that each time $\Phi(i,l) = 0$, the algorithm is at the optimum, however, the opposite is not true: the optimum can be reached, but not all states become learned. This does not hurt anything: the additive drift theorem gives an upper bound on the number of iterations until the condition that the optimum is reached *and* all the states are learned, which, in turn, is an upper bound on the actual running time of the algorithm.

We can treat $\Phi(i,l)$ as a sum of two functions, $\Phi_1(i) = \sum_{t=i}^{N-1} \frac{N}{N-t}$ and $\Phi_2(l) = CN \sum_{t=0}^{N-1} \frac{1-l(t)}{N-t}$. As Φ_1 is upwards convex, from Jensen's inequality it follows that $\Phi_1(i) - E(\Phi_1(i')) \geq \Phi_1(i) - \Phi_1(E(i'))$, if Φ_1 is extended to non-integer arguments by linear interpolation.

5.1 Drift from a Learned State

In a learned state the situation resembles how $(1+1)$-EA works on ONEMAX [11]. In this case, the learn indicator l does not change, so drift of Φ_2 is zero. The lower bound on $E(i')$ is (using Theorem 5):

$$E(i') = i + \sum_{j=0}^{N} (j-i)P_L^{i,j} = i + \sum_{j=i+1}^{N} (j-i)P^{i,j} \geq i + \frac{N-i}{N-1}\left(\frac{N-1}{N}\right)^{i+1}.$$

The lower bound on $E(i') - i$ is at most one. The drift of Φ_1 (and thus Φ) is:

$$\Phi_1(i) - E(\Phi_1(i')) \geq \Phi_1(i) - \Phi_1(E(i')) \geq \frac{N}{N-i}\frac{N-i}{N-1}\left(\frac{N-1}{N}\right)^{i+1} \geq e^{-1}.$$

5.2 Drift from an Unlearned State

In an unlearned state, the expected drift of $\Phi_2(l)$ can be estimated as the reward for learning the current state i multiplied by the probability the algorithm leaves the state i (using Theorems 2 and 3). $\Phi_2(l) - E(\Phi_2(l))$ is at least:

$$C\frac{N}{N-i}\sum_{j\neq i}\frac{P^{i,j}}{2} \geq \frac{C}{2}\frac{N}{N-i}\left(\left(\frac{N-1}{N}\right)^{N-i} + \left(\frac{N-1}{N}\right)^{i} - 2\left(\frac{N-1}{N}\right)^{N}\right).$$

The lower bound on $E(i')$ is (using Theorems 4 and 5):

$$E(i') = i + \sum_{j=0}^{N} P_U^{i,j}(j-i) = i + \frac{1}{2}\sum_{j\neq i} P^{i,j}(j-i)$$

$$= i + \frac{1}{2}\left(\sum_{j=i+1}^{N} P^{i,j}(j-i) - \sum_{j=0}^{i-1} P^{i,j}(i-j)\right)$$

$$\geq i + \frac{1}{2}\left(\frac{N-i}{N-1}\left(\frac{N-1}{N}\right)^{i+1} - R\frac{i}{N-1}\left(\frac{N-1}{N}\right)^{N-i+1}\right).$$

As $R \leq \frac{8}{7}$, the lower bound on $E(i') - i$ is at most $-\frac{4}{7}$, so the drift of Φ_1 is:

$$\Phi_1(i) - E(\Phi_1(i')) \geq \Phi_1(i) - \Phi_1(E(i'))$$

$$\geq \frac{1}{2}\frac{N}{N-i+1}\left(\frac{N-i}{N-1}\left(\frac{N-1}{N}\right)^{i+1} - \frac{Ri}{N-1}\left(\frac{N-1}{N}\right)^{N-i+1}\right)$$

$$= \frac{1}{2}\frac{N-i}{N-i+1}\left(\frac{N-1}{N}\right)^{i} - \frac{R}{2}\frac{i}{N-i+1}\left(\frac{N-1}{N}\right)^{N-i}.$$

The total value of Φ, namely, $D = \Phi(i,l) - E(\Phi(i',l'))$, is bounded from below by sum of drifts for Φ_1 and Φ_2:

$$D \geq \frac{C}{2} \frac{N}{N-i} \left(\left(\frac{N-1}{N} \right)^{N-i} + \left(\frac{N-1}{N} \right)^{i} - 2 \left(\frac{N-1}{N} \right)^{N} \right)$$

$$+ \frac{1}{2} \frac{N}{N-i+1} \left(\frac{N-i}{N} \left(\frac{N-1}{N} \right)^{i} - \frac{Ri}{N} \left(\frac{N-1}{N} \right)^{N-i} \right)$$

$$\geq \frac{C}{2} \frac{N}{N-i+1} \left(\left(\frac{N-1}{N} \right)^{N-i} + \left(\frac{N-1}{N} \right)^{i} - 2 \left(\frac{N-1}{N} \right)^{N} \right)$$

$$+ \frac{1}{2} \frac{N}{N-i+1} \left(\frac{N-i}{N} \left(\frac{N-1}{N} \right)^{i} - \frac{Ri}{N} \left(\frac{N-1}{N} \right)^{N-i} \right)$$

$$= \frac{\left(\frac{N-1}{N} \right)^{N-i} (CN - Ri) + \left(\frac{N-1}{N} \right)^{i} (CN + N - i) - \left(\frac{N-1}{N} \right)^{N} (2CN)}{2(N-i+1)}.$$

If $C \geq R$, then $CN - Ri$ is positive. As $\left(\frac{N-1}{N} \right)^{x}$ is convex downwards, we can use Jensen's inequality to simplify a part of the latter expression, which we call G:

$$G = \left(\frac{N-1}{N} \right)^{N-i} (CN - Ri) + \left(\frac{N-1}{N} \right)^{i} (CN + N - i)$$

$$= \frac{\left(\frac{N-1}{N} \right)^{N-i} \frac{CN-Ri}{(2C+1)N-(R+1)i} + \left(\frac{N-1}{N} \right)^{i} \frac{CN+N-i}{(2C+1)N-(R+1)i}}{((2C+1)N - (R+1)i)^{-1}}$$

$$\geq ((2C+1)N - (R+1)i) \left(\frac{N-1}{N} \right)^{\frac{(N-i)(CN-Ri)+i(CN+N-i)}{(2C+1)N-(R+1)i}}.$$

To find a lower bound on G one needs to find an upper bound on the exponent in the expression above, assuming that $0 \leq i \leq N$. Recall that $R = \frac{8}{7}$, which makes the exponent be equal to $\frac{i^2 - iN + 7CN^2}{14CN + 7N - 15i}$. The derivative by i of this exponent have no roots in $[0; N]$ at least when $C \geq 1$. The value for $i = 0$ is $N \frac{C}{2C+1}$, and for $i = N$ it is $N \frac{7C}{14C-8}$, the second one is the biggest of two. We continue with the lower bound on D:

$$D \geq \frac{((2C+1)N - (R+1)i) \left(\frac{N-1}{N} \right)^{N \frac{7C}{14C-8}} - \left(\frac{N-1}{N} \right)^{N} (2CN)}{2(N-i+1)}$$

$$= \frac{\left(\left(1 + 2C \left(1 - \left(\frac{N-1}{N} \right)^{N \frac{7C-8}{14C-8}} \right) \right) N - (R+1)i \right)}{2(N-i+1)} \left(\frac{N-1}{N} \right)^{N \frac{7C}{14C-8}}.$$

If we choose C such that $1 + 2C \left(1 - \left(\frac{N-1}{N} \right)^{N \frac{7C-8}{14C-8}} \right) > R + 1$, then starting from some N the fraction will be greater than one. For these needs we approximate $\left(\frac{N-1}{N} \right)^{N}$ by e^{-1}. This problem can be reduced to finding a minimum C such that $1 - \frac{4}{7C} > e^{\frac{7C-8}{8-14C}}$. This can be done by a binary search which yields $C = 2.115188060 \ldots \approx 2.12$. Consequently, when C is at least this large, the

Table 1. Experiment results. $C \approx 2.12$ is a constant which was proven.

N	Average FF calls		Average false	$2eN \log N$	$(1+C)eN \log N$	Ratio to
	$\gamma = 1/N$	$\gamma = 1$	queries, $\gamma = 1$			$\gamma = 1/N$
$1 \cdot 10^1$	$9.892 \cdot 10^1$	$9.648 \cdot 10^1$	$3.000 \cdot 10^{-2}$	$1.252 \cdot 10^2$	$1.950 \cdot 10^2$	1.97
$3 \cdot 10^1$	$4.673 \cdot 10^2$	$4.855 \cdot 10^2$	$1.900 \cdot 10^{-1}$	$5.547 \cdot 10^2$	$8.640 \cdot 10^2$	1.85
$1 \cdot 10^2$	$2.382 \cdot 10^3$	$2.389 \cdot 10^3$	$7.800 \cdot 10^{-1}$	$2.504 \cdot 10^3$	$3.900 \cdot 10^3$	1.64
$3 \cdot 10^2$	$9.340 \cdot 10^3$	$9.335 \cdot 10^3$	$1.690 \cdot 10^0$	$9.303 \cdot 10^3$	$1.449 \cdot 10^4$	1.55
$1 \cdot 10^3$	$3.910 \cdot 10^4$	$3.925 \cdot 10^4$	$8.280 \cdot 10^0$	$3.755 \cdot 10^4$	$5.849 \cdot 10^4$	1.50
$3 \cdot 10^3$	$1.389 \cdot 10^5$	$1.441 \cdot 10^5$	$2.414 \cdot 10^1$	$1.306 \cdot 10^5$	$2.034 \cdot 10^5$	1.46
$1 \cdot 10^4$	$5.585 \cdot 10^5$	$5.461 \cdot 10^5$	$7.443 \cdot 10^1$	$5.007 \cdot 10^5$	$7.799 \cdot 10^5$	1.40
$3 \cdot 10^4$	$1.882 \cdot 10^6$	$1.901 \cdot 10^6$	$2.330 \cdot 10^2$	$1.681 \cdot 10^6$	$2.619 \cdot 10^6$	1.39
$1 \cdot 10^5$	$7.225 \cdot 10^6$	$7.108 \cdot 10^6$	$7.780 \cdot 10^2$	$6.259 \cdot 10^6$	$9.749 \cdot 10^6$	1.35
$3 \cdot 10^5$	$2.376 \cdot 10^7$	$2.376 \cdot 10^7$	$2.325 \cdot 10^3$	$2.057 \cdot 10^7$	$3.204 \cdot 10^7$	1.35
$1 \cdot 10^6$	$8.726 \cdot 10^7$	$8.648 \cdot 10^7$	$7.632 \cdot 10^3$	$7.511 \cdot 10^7$	$1.170 \cdot 10^8$	1.34

drift from an unlearned state is at least $\left(\frac{N-1}{N}\right)^{N \frac{7C}{14C-8}} \approx \left(\frac{N-1}{N}\right)^{0.6845N} \geq e^{-1}$. Together with all previous analysis, this proves the following theorem:

Theorem 6. *The expected running time of the* $(1 + 1)$ *evolutionary algorithm controlled by greedy Q-learning on the* ONEMAX+ZEROMAX *problem with a value of the discount factor* $\gamma < 1/(N - 1)$ *is at most:*

$$(1 + C)eN \log N \approx 3.12eN \log N.$$

6   Experimental Evaluation

We conducted experimental evaluation on big problem sizes to see how precise our estimations are. Table 1 presents the results. For each N, there were 100 runs, and the numbers of fitness function calls were averaged. We tested two values of the discount factor γ, namely $1/N$ and 1. For the latter value, we additionally track the number of situations when Q values were tuned wrong and ZEROMAX was chosen intentionally (the "Average false queries" column). Additionally, we track the value of $2eN \log N$, which was the common belief for the "right" expression on the algorithm's runtime.

First, it turned out that for large N the algorithm needs more than $2eN \log N$ iterations to find an optimum, which was surprising. Second, the runtimes for different values of γ seem to be the same. The false queries column provides an insight for this phenomenon: for $\gamma = 1$, the number of mismatches with the learning lemma seems to be $\Theta(N)$ with a constant approximately equal to $7.6 \cdot 10^{-3}$. We have no proof for this yet, but it seems that the probability of "learning the wrong way" is very small and in most cases the algorithm can later "self-heal" from wrong decisions. Finally, our estimation appears to be quite a good one, as our estimations are only 35 % pessimistic for large N.

7   Conclusion

We presented a proof that the $(1+1)$ evolutionary algorithm, when controlled by the EA+RL method which uses Q-learning with a greedy exploration strategy, small values of the discount factor $\gamma < 1/(N - 1)$, difference in target fitness function as a reward and reinforcement learning states determined by the target fitness function, solves the ONEMAX+ZEROMAX problem in $O(N \log N)$—more precisely, in at most $3.12eN \log N$ fitness function calls in expectation.

Experiments show that early thoughts on the actual expected running time, $2eN \log N$, are wrong for $N \geq 10^5$. The current upper bound seems to be only 35 % worse than the "real life". What is more, the influence of big values of γ is shown to be negligible. We hope that these results will show the way for proving bounds on the running time for the EA+RL method on more complex problems.

This work was partially financially supported by the Government of Russian Federation, Grant 074-U01.

References

1. Brockhoff, D., Friedrich, T., Hebbinghaus, N., Klein, C., Neumann, F., Zitzler, E.: On the effects of adding objectives to plateau functions. Trans. Evol. Comput. **13**(3), 591–603 (2009)
2. Buzdalov, M., Buzdalova, A., Shalyto, A.: A first step towards the runtime analysis of evolutionary algorithm adjusted with reinforcement learning. In: Proceedings of the International Conference on Machine Learning and Applications, vol. 1, pp. 203–208. IEEE Computer Society (2013)
3. Buzdalova, A., Buzdalov, M.: Increasing efficiency of evolutionary algorithms by choosing between auxiliary fitness functions with reinforcement learning. In: Proceedings of the International Conference on Machine Learning and Applications, vol. 1, pp. 150–155 (2012)
4. Hajek, B.: Hitting-time and occupation-time bounds implied by drift analysis with applications. Adv. Appl. Probab. **14**(3), 502–525 (1982)
5. Handl, J., Lovell, S.C., Knowles, J.D.: Multiobjectivization by decomposition of scalar cost functions. In: Rudolph, G., Jansen, T., Lucas, S., Poloni, C., Beume, N. (eds.) PPSN 2008. LNCS, vol. 5199, pp. 31–40. Springer, Heidelberg (2008)
6. Jensen, M.T.: Helper-objectives: using multi-objective evolutionary algorithms for single-objective optimisation: evolutionary computation combinatorial optimization. J. Math. Model. Algorithms **3**(4), 323–347 (2004)
7. Knowles, J.D., Watson, R.A., Corne, D.W.: Reducing local optima in single-objective problems by multi-objectivization. In: Zitzler, E., Deb, K., Thiele, L., Coello Coello, C.A., Corne, D.W. (eds.) EMO 2001. LNCS, vol. 1993, pp. 269–283. Springer, Heidelberg (2001)
8. Lochtefeld, D.F., Ciarallo, F.W.: Helper-objective optimization strategies for the job-shop scheduling problem. Appl. Soft Comput. **11**(6), 4161–4174 (2011)
9. Neumann, F., Wegener, I.: Can single-objective optimization profit from multiobjective optimization? In: Knowles, J., Corne, D., Deb, K., Chair, D.R. (eds.) Multiobjective Problem Solving from Nature. Natural Computing Series, pp. 115–130. Springer, Heidelberg (2008)

10. Sutton, R.S., Barto, A.G.: Reinforcement Learning: An Introduction. MIT Press, Cambridge (1998)
11. Witt, C.: Optimizing linear functions with randomized search heuristics - the robustness of mutation. In: Proceedings of the 29th Annual Symposium on Theoretical Aspects of Computer Science, pp. 420–431 (2012)

The New Memetic Algorithm *HEAD* for Graph Coloring: An Easy Way for Managing Diversity

Laurent Moalic[1](✉) and Alexandre Gondran[2]

[1] UTBM, OPERA, University of Technology of Belfort-Montbéliard,
90010 Belfort Cedex, France
laurent.moalic@utbm.fr
[2] MAIAA, ÉNAC, French Civil Aviation University, 31055 Toulouse Cedex 4, France
alexandre.gondran@enac.fr

Abstract. This paper presents an effective memetic approach *HEAD* designed for coloring difficult graphs. In this algorithm a powerful tabu search is used inside a very specific population of individuals. Indeed, the main characteristic of *HEAD* is to work with a population of only two individuals. This provides a very simple algorithm with neither selection operator nor replacement strategy. Because of its simplicity, *HEAD* allows an easy way for managing the diversity. We focus this work on the impact of this diversity management on well-studied graphs of the DIMACS challenge benchmarks, known to be very difficult to solve. A detailed analysis is provided for three graphs on which *HEAD* finds a legal coloring with less colors than reference algorithms: DSJC500.5 with 47 colors, DSJC1000.5 with 82 colors and flat1000_76_0 with 81 colors. The analysis performed in this work will allow to improve *HEAD* efficiency in terms of computation time and maybe to decrease the number of needed colors for other graphs.

1 Introduction

Graph coloring consists of assigning a color to each vertex of a given graph $G = (V, E)$, such that all pairs of adjacent vertices (linked by an edge) are assigned different colors. Such coloring is said legal. V is known as the vertices set and E edges set. The Graph Coloring Problem (GCP) is to find, for a given graph, the minimum number of colors that provides a legal coloring. This minimum number is the *chromatic number* $\chi(G)$. GCP is a very famous and important because lots of applied problems can be modeled as a GCP: frequency assignment problem [1], timetabling problem [2], scheduling problem [3], fly level allocation problem [4], and many others. Since GCP is NP-hard [5], exact algorithms can be used only with easy or small graphs, heuristic approaches are required with more complex graphs.

The decision problem linked to GCP is the k-coloring problem. It consists on determining if it is possible to to find a legal coloring with k colors. This is a NP-complete problem [5] for $k > 2$.

© Springer International Publishing Switzerland 2015
G. Ochoa and F. Chicano (Eds.): EvoCOP 2015, LNCS 9026, pp. 173–183, 2015.
DOI: 10.1007/978-3-319-16468-7_15

Most of computational optimization techniques have been applied to solve GCP. We will present a short overview of these techniques grouped into five categories.

The first approaches are constructive methods. One can find in this category greedy methods whose DSATUR [6] and RLF [7] are the most well-known, and can be used to quickly provide an upper bound.

Many exact methods have been developed to solve GCP for small graphs: branch and bound [8], backtracking [9], constraint programming [10,11], linear programming [12], column generation [10,13].

Local searches such as hill-climbing method [14], simulated annealing [15], tabu search [16] try to improve an initial solution by local moves. These approaches were the first ones able to find good coloring for large graphs. Some improvements have been brought successfully by using in the same algorithm different local strategies, such as in variable neighborhood search [17] or in variable space search [18] or [19].

And finally, some population-based approaches have been developed with interesting results. Evolutionary Algorithms [20,21], ant colony optimization [22], particle swarm algorithm, are some of them. They work with several colorings that can interact together with a crossover operator, a shared memory, or a repulsion/attraction operator.

All these methods provide interesting results with more or less efficiency. But the best known algorithms which are able to solve the most difficult GCP are hybridizations of previous algorithms. In many cases they use a powerful local search inside a population-based algorithm. The memetic Algorithms [23–25] and quantum annealing [26–28] provide up to now the most interesting approaches.

In this paper we present $HEAD$ (Hybrid Evolutionary Algorithm in Duet), a new and very simple hybrid algorithm providing the best results on most of the DIMACS graphs [29]. These good results are allowed thanks to a simple way for managing the diversity. Indeed, the control of diversity is one of the key aspect for metaheuristics. We focus our analysis in this article on this management for three difficult graphs. For these graphs $HEAD$ allows to decrease the needed number of colors: DSJC500.5 (solved with 47 colors), DSJC1000.5 (with 82 colors) and flat1000_76_0 (with 81 colors).

This article is organized as follows. The hybrid algorithm $HEAD$ is described in Sect. 2. The experimental results are presented in Sect. 3 and the analysis of the diversification impact are detailed in Sect. 4. Finally we draw the conclusions of this study and the future works to do in Sect. 5.

2   Algorithm: $HEAD$

Many powerful algorithms are based on the hybridization of other more simple components. When an hybrid algorithm uses a local search inside a population based approach, it is called Memetic Algorithm (MA) or Genetic Local Search (GLS). In a MA the classical mutation is replaced by a local search for providing a better intensification. $HEAD$ belong to this family. It can be seen as an extension of the Hybrid Evolutionary Algorithm (HEA) of Galinier and Hao [23] known to be one of the best algorithms for solving GCP. Indeed HEA produces most of best results since 1999 on DIMACS benchmark [30], especially for very difficult graphs such as those detailed in this article (see Table 1).

HEA and $HEAD$ share the same two main elements: the local search strategy and the crossover. The local search used in both is an improvement of the TabuCol of [16]. This is a powerful tabu local search able to provide good results, even if used alone (not combined with a population based approach). The common crossover is called Greedy Partition Crossover (GPX). Its main characteristic is to build a child with the biggest color classes of each parent. Both TabuCol and GPX work with a k-fixed penalty strategy that accept no proper solutions with conflicts. The aim of HEA and $HEAD$ is to find a coloring which minimize the number of conflicting edges.

The population involved in memetic algorithms brings the diversity while local search is used for intensifying the solutions. The greater the population size, the greater the diversification. That is why in HEA a population size of 10 is usually selected. In $HEAD$ we propose a new easy way to manage diversification. To achieve that objective the population size is reduced to only 2 individuals. By this method $HEAD$ can be seen as two parallel TabuCol. One of the main drawbacks of such a simple population is that sometimes it doesn't provide enough diversification. Indeed after a given number of generation, the two parents are quite similar. That is why diversification must be reintroduce, which is performed by applying an elite strategy. For a cycle (a cycle is defined as a given number of generations equals to $cycleSize$) the best individual (known as elite) is recorded. After each cycle the elite from the previous cycle is reintroduced to replace one of the two population members. This elite mechanism can be considered as selection and replacement strategies very easy to manage.

We present the pseudo-code of $HEAD$ in Algorithm 1. It uses few parameters: $Iter_{TC}$, the number of iterations performed by the TabuCol algorithm, $cycleSize$, the number of generations inside one cycle and ϵ for the stop condition.

3   Results

The aim of this paper is not to present results of best coloring for numerous graphs, but it is to provide an analysis of how the results were found. Therefore, the results presented come from [29] and are obtained on graphs from the second DIMACS challenge of 1992-1993 [30]. It is to date the most widely used benchmark for solving the graph coloring problem.

Two main types of graphs of DIMACS benchmark are taken into account: DSJC and FLAT, which are randomly or quasi-randomly generated. DSJC$n.d$ graphs are graphs with n vertices which each vertex is connected to an average of $n \times d$ vertices; d is the graph density. The chromatic number of these graphs is unknown. FLAT graphs have another structure. They are built for a known chromatic number. The flatn_χ graph has n vertices and χ is the chromatic number.

$HEAD$ was programmed in standard C++. The results presented in this section were obtained on a computer with an Intel Xeon 3.10GHz processor - 4 cores

---

Algorithm 1. $HEAD$: HEA in Duet for k-coloring problem

1: **Input:** a graph: $G(V, E)$ with $n = |V|$, a number of colors: k , an crossover: GPX, a local search: TabuCol, some parameters: $Iter_{TC}$, $cycleSize = 10$, $\epsilon = 0.01$.
2: **Output:** the best configuration found
3: $p_1, p_2, elite_1, elite_2 \leftarrow$ init() {initialize with random colorings}
4: $generation \leftarrow 0$
5: **repeat**
6:      $c_1 \leftarrow$ GPX(p_1, p_2)
7:      $c_2 \leftarrow$ GPX(p_2, p_1)
8:      $p_1 \leftarrow$ TabuCol($c_1, Iter_{TC}$)
9:      $p_2 \leftarrow$ TabuCol($c_2, Iter_{TC}$)
10:     $elite_1 \leftarrow$ saveBest($p_1, p_2, elite_1$)
11:     $best \leftarrow$ saveBest($elite_1, best$)
12:     **if** $generation\%cycleSize = 0$ or $d_H(p_1, p_2) \leq n\epsilon$ **then**
13:         $p_1 \leftarrow elite_2$
14:         $elite_2 \leftarrow elite_1$
15:         $elite_1 \leftarrow$ init()
16:     **end if**
17:     $generation + +$
18: **until** $stop\_condition : d_H(p_1, p_2) \leq n\epsilon$
19: **return** $best$

---

and 8GB of RAM. Almost all of the computation time is spent performing the tabu search (lines 8 and 9 of Algorithm 1). It is possible and easy to parallelize both tabu searches by using a multi-core processor architecture. This is what we have done using the OpenMP API (Open Multi-Processing). The execution times in the following table are CPU time. Thus, when we give an execution time of 30 min, the needed time is actually close to 15 min using two processing cores.

Table 1 presents results of the principal methods known to date. For each graph, it indicates the lowest number of colors found by each algorithm. The most recent algorithm, QA-col (Quantum Annealing for graph coloring [28]), provides the best results but is based on a cluster of PCs using 10 processing cores simultaneously. For detail on those approaches refer to [29].

Table 2 presents results obtained with $HEAD$. The first important thing is that $HEAD$ find solutions with fewer colors than all the best known methods. The column $Iter_{TC}$ gives the number of iterations of TabuCol algorithm (it is the stop criteria of TabuCol). The column **Success** evaluates the robustness of the method, it gives the success rate: success_runs/total_runs. A success run is one which finds a legal k-coloring. The average number of generations or crossovers done during one success run is given by **Gene** value. Then the total average number of iterations of TabuCol preformed during $HEAD$ is **Iter**= $LS \times Gene \times 2$. The column **Time** gives the average CPU time in minutes of success runs. For all these results a cycle size of 10 is chosen. We will see in Sect. 4.2 that it is possible to tune this parameter for improving the results.

Table 1. Comparison between $HEAD$ and references algorithms

Graphs	2014 $HEAD$	2008 TabuCol	1999 HEA	2008 AmaCol	2010 MACOL	2011 EXTRACOL	2012 QA-col
DSJC250.5	28	28	28	28	28	-	28
DSJC500.1	12	13	-	12	12	-	12
DSJC500.5	47	50	48	48	48	-	47
DSJC500.9	126	127	-	126	126	-	126
DSJC1000.1	20	20	20	20	20	20	20
DSJC1000.5	82	89	83	84	83	83	82
DSJC1000.9	222	227	224	224	223	222	222
r250.5	65	-	-	-	65	-	65
r1000.1c	98	-	-	-	98	101	98
r1000.5	245	-	-	-	245	250	-
DSJR500.1c	85	85	-	86	85	-	85
le450_25c	25	26	26	26	25	-	25
le450_25d	25	26	-	26	25	-	25
flat300_28_0	31	31	31	31	29	-	31
flat1000_50_0	50	50	-	50	50	50	-
flat1000_60_0	60	60	-	60	60	60	-
flat1000_76_0	81	88	83	84	82	82	81

4   Analysis of the Diversity

As we have presented previously, diversification plays a key role in metaheuristics. In memetic algorithms, one of the major diversification operator is the crossover. The distance between the parents provides the level of diversity. That is why it is so important to control precisely this distance between the individuals: controlling the distance between individuals mean controlling the impact of diversification. With only two individuals, $HEAD$ allows a very simple control of this diversification. One can consider that such an algorithm is a two parallel local searches in which we introduce a diversification operator with the crossover.

4.1   Relationship Between Distance and Fitness Values

The fitness analysis is the first classical way to provide information on how an algorithm works. Here we define the fitness value as the number of edges in conflict. Two main phases of the algorithm $HEAD$ can be identified. The first one is the local search in which a solution is improved by small moves. The second one is the global search performed by the crossover. We are interested here by the global search progress for the specific graph DSJC500.5 with 47 colors.

A fitness value is determined at the end of each generation for both individuals. Figure 1 shows the results with the green and red lines for the fitness values of each individual. In the same view is displayed the corresponding Hamming proximity p_H between both individuals (blue line). It is computed as the

Table 2. Results of $HEAD$ algorithm with the indication of CPU time in minutes

Instances	k	Iter$_{TC}$	Success	Iter	Gene	Time
Dsjc250.5	28	6000	20/20	0.9×10^6	77	0.03
dsjc500.1	12	4000	20/20	3.8×10^6	483	0.1
dsjc500.5	47	8000	2/10000	24×10^6	1517	1.7
	48	8000	20/20	7.6×10^6	479	0.5
dsjc500.9	126	15000	13/20	29×10^6	970	2
dsjc1000.1	20	3000	20/20	3.4×10^6	567	0.2
dsjc1000.5	82	40000	1/20	548×10^6	6854	48
	83	40000	20/20	96×10^6	1200	8
dsjc1000.9	222	50000	1/20	1.4×10^9	14208	174
	223	30000	19/20	126×10^6	2107	16
r250.5	65	2000	20/20	1.0×10^9	255287	79
r1000.1c	98	25000	20/20	4.7×10^6	95	0.3
r1000.5	245	360000	20/20	3.3×10^9	4628	176
DSJR500.1c	85	400	20/20	0.4×10^6	534	0.02
le450_25c	25	27000000	20/20	2.2×10^9	42	67
	26	70000	20/20	0.17×10^6	1.4	0.01
le450_25d	25	300000	20/20	549×10^6	916	14
flat300_28_0	31	4000	20/20	0.9×10^6	120	0.04
flat1000_50_0	50	130000	20/20	1.1×10^6	5	0.3
flat1000_60_0	60	130000	20/20	2.2×10^6	9	0.5
flat1000_76_0	81	40000	1/20	716×10^6	8961	96
	82	40000	20/20	84×10^6	1052	8

complement of the Hamming distance $d_H : p_H = n - d_H$ where n is the number of vertices of the graph. We reuse the definition given by [24] of the distance between two k-colorings. The distance (called Hamming distance) between two k-colorings c_1 and c_2 is defined as the least number of 1-move steps (i.e. a color change of one vertex) for transforming c_1 to c_2. Of course this distance has to be independent of the permutation of the color classes, then before counting the number of 1-moves, we have to match each color class of c_1 with the nearest color class of c_2. This problem is a maximum weighted bipartite matching if we consider each color class of c_1 and c_2 as the vertices of a bipartite graph; an edge links a color class of c_1 with a color class of c_2 with an associated value corresponding to the number of vertices shared by those classes. Then the distance is $d_H(c_1, c_2) = n - p_H$ where n is the number of vertices of the initial graph and p_H the result of the matching (called Hamming proximity); i.e. the maximal total number of sharing vertices in the same class for c_1 and c_2.

Three main stages can be identified in Fig. 1. The first one for generations 0 to 200 provides a global improvement of the individuals (intensification part). From generations 200 to 700 the algorithm provides an exploration of the search space with a number of conflicts between 5 to 20. The last stage, at the very end of the algorithm, provides a new intensification of the individuals until finding the

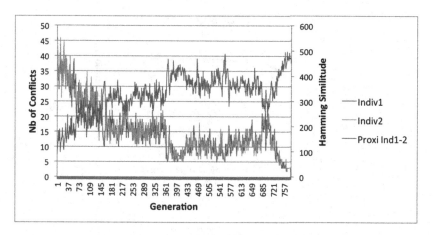

Fig. 1. Fitness values recorded after each local search (generation) for one run finding a legal 47-coloring in DSJC500.5 (500 vertices, 62624 edges)(Color figure online)

legal solution. The fitness shape shows that until the very end of the algorithm, it is not possible to predict whether a valid coloring will be found.

The red and green lines of Fig. 1 show that the individuals are still very similar in terms of fitness value: the two individuals are good at the same time. The Hamming proximity shows an important aspect of the algorithm. It reveals that the number of conflicts is directly linked to the proximity between the individuals. The closer the individuals are, the better they are. When the algorithm find the legal coloration, both individuals are very close. Figure 1 allows to understand this link existing between fitness value and proximity of individuals. Now the point is to control this proximity between individuals in order to find a legal coloring before having two individuals too close together.

4.2 Control of the Diversity with the Swapping Frequency

As presented before, *HEAD* can be seen as a two individuals local search algorithm with a specific diversification operator: the crossover. More precisely, one can identify three levels of diversification in the *HEAD* algorithm:

- the first level is brought by the Tabu tenure (small impact)
- the second level is brought by the crossover (medium impact)
- the third level is brought by the elite mechanism (great impact)

The first two levels have been much studied in the literature and have shown their efficiency for solving the k-coloring problem. For the first level we consider the standard parameters for TabuCol as defined in [23]. The Tabu tenure is defined by the values $L = 10$ and $\alpha = 0.6$ which are known to give good results. It could be tuned thanks to the considered graph to provide even better results. For the second level the crossover is the standard GXP Algorithm (see [23]). Some modifications of GPX is done in [24, 29, 31].

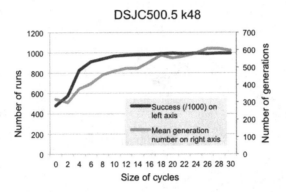

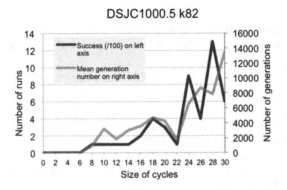

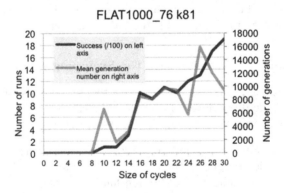

Fig. 2. Number of successes depending on the swapping frequency for three graphs: DSJC500.5 (2-a), DSJC1000.5 (2-b) and FLAT1000.5 (2-c)

Of course it is possible to adapt the diversification rate by changing the Tabu parameters or the algorithms themselves. But because they have been yet very studied we will focus our analysis on the third level of diversification. This is the only parameter specific to *HEAD* and the one responsible to the good performances reached by *HEAD*.

The swapping operation consists of reintroducing the elite solution (the best solution from the previous cycle) to the population. In *HEAD*, two types of elements can make a swapping operation. The first one is the number of generation. At the end of each cycle (i.e. *cycleSize* generations) the previous elite is reintroduced. The second one is the proximity between the two individuals. As soon as it is higher than a given rate (99 %), the crossover has almost no effect and cannot bring diversity: a swapping operation is performed.

Figure. 2 show the impact of the cycle size on the ability of finding valid colorings. Each figure corresponds to a test on a specific graph (DSJC500.5 for Fig. 2-a, DSJC1000.5 for Fig. 2-b and FLAT1000.5 for Fig. 2-c). In Fig. 2-a (respectively Figs. 2-b-c) we run *HEAD* 1000 times (respectively 100 times for Figs. 2-b-c) to solve the k-coloring problem with $k = 48$ for DSJC500.5 (respectively with $k = 82$ for DSJC1000.5 and $k = 81$ for FLAT1000_76). In x-axis, we increase the cycle size. In left y-axis we count the number of successful runs, in right y-axis we count the average number of generations needed for get a legal coloring. The second aspect corresponds to the time efficiency of the algorithm. Several interesting results can be noticed. First we will focus on the cycle size equal to 0 (no swapping). It is the case of HEA with a population of two individuals, that we call HEAD' (HEA in Duet without elite). Neither DSJC1000.5 nor FLAT1000.5 can be legally colored with respectively 82 and 81 colors. Only DSJC500.5 can be solved (478 legal k-coloring for 1000 runs) with 48 colors. That clearly means that without an elite mechanism the algorithm converge too fast and the amount of diversification is not enough. The algorithm doesn't allow to escape from local optima.

On the other hand, when a success occurs, the number of generations for finding the solution is quite low. Only 317 generations in mean are necessary to find the solution for the graph DSJC500.5.

When the cycle size increases, generally the number of successes increase too. But It seems interesting to notice that for DSJC500.5, the success number increase more quickly than the generations needed. Moreover, for the two other graphs, a swapping rate of 8 is required for finding legal solutions. A value of 28 is more appropriate.

This figures show clearly the role of the cycle size to control the diversity (the balance between the time efficiency of the algorithm and its quality efficiency) and how much it allows to improve significantly HEAD' for solving the k-coloring problem.

5   Conclusion

We have presented an effective memetic algorithm for the graph coloring problem, called *HEAD*. It is a variation of the Galinier and Hao [23]'s algorithm with only two candidate solutions. *HEAD* produces very good results on a set of challenging DIMACS graphs. We have focused our analysis on the control of diversification in our memetic algorithm which has been detailed for three specific graphs of the DIMACS benchmarks, where $HEAD$ improve significantly the results achieved by reference algorithms. We have shown that the Hamming distance between the two individuals of the population is strongly correlated to the fitness value of the individuals.

The analyses performed in this work have demonstrated the interest of working with only two individuals. Moreover, they provide us a better understanding of $HEAD$ and define some important tools to improve even more this algorithm.

Acknowledgements. The second author gratefully acknowledge financial support under grant ANR 12-JS02-009-01 "ATOMIC".

References

1. Aardal, K., Hoesel, S., Koster, A., Mannino, C., Sassano, A.: Models and solution techniques for frequency assignment problems. Q. J. Belg. Fr. Ital. Oper. Res. Soc. **1**(4), 261–317 (2003). doi:10.1007/s10288-003-0022-6
2. Wood, D.C.: A technique for coloring a graph applicable to large-scale timetabling problems. Comput. J. **12**, 317–322 (1969)
3. Zufferey, N., Amstutz, P., Giaccari, P.: Graph colouring approaches for a satellite range scheduling problem. J. Sched. **11**(4), 263–277 (2008)
4. Barnier, N., Brisset, P.: Graph coloring for air traffic flow management. Ann. Oper. Res. **130**(1–4), 163–178 (2004). doi:10.1023/B:ANOR.0000032574.01332.98
5. Karp, R.: Reducibility among combinatorial problems. In: Miller, R.E., Thatcher, J.W. (eds.) Complexity of Computer Computations, pp. 85–103. Plenum Press, New York (1972)
6. Brélaz, D.: New methods to color the vertices of a graph. Commun. ACM **22**(4), 251–256 (1979)
7. Leighton, F.T.: A graph coloring algorithm for large scheduling problems. J. Res. Natl. Bur. Stan. **84**(6), 489–506 (1979)
8. Glover, F., Parker, M., Ryan, J.: Coloring by tabu branch and bound. DIMACS Ser. Discrete Math. Theor. Comput. Sci. **26**, 285–307 (1996)
9. Zykov, A.A.: On some properties of linear complexes. Mat. Sb. (N.S.) **24**(66:2), 163–188 (1949)
10. Gualandi, S., Malucelli, F.: Exact solution of graph coloring problems via constraint programming and column generation. INFORMS J. Comput. **24**(1), 81–100 (2012). doi:10.1287/ijoc.1100.0436
11. Caramia, M., Dell'Olmo, P.: Constraint propagation in graph coloring. J. Heuristics **8**(1), 83–107 (2002)
12. Schindl, D.: Graph coloring and linear programming, presentation at First Joint Operations Research Days, Ecole Polytechnique Fédérale de Lausanne (EPFL), available on line (last visited June 2005) (July 2003). http://roso.epfl.ch/ibm/jord03.html
13. Mehrotra, A., Trick, M.A.: A column generation approach for graph coloring. INFORMS J. Comput. **8**(4), 344–354 (1996)
14. Lewis, R.: A general-purpose hill-climbing method for order independent minimum gr ouping problems: a case study in graph colouring and bin packing. Comput. Oper. Res. **36**(7), 2295–2310 (2009). doi:10.1016/j.cor.2008.09.004. http://www.sciencedirect.com/science/article/B6VC5-4TGHNJ4-1/2/1040b5ca8 ef6fc2ddf012f32f3de9cb5

15. Johnson, D.S., Aragon, C.R., McGeoch, L.A., Schevon, C.: Optimization by simulated annealing: An experimental evaluation; part II, graph coloring and number partitioning. Oper. Res. **39**(3), 378–406 (1991)
16. Hertz, A., de Werra, D.: Using tabu search techniques for graph coloring. Computing **39**(4), 345–351 (1987)
17. Avanthay, C., Hertz, A., Zufferey, N.: A variable neighborhood search for graph coloring. Eur. J. Oper. Res **151**(2), 379–388 (2003). Elsevier
18. Hertz, A., Plumettaz, M., Zufferey, N.: Variable space search for graph coloring. Discret. Appl. Math. **156**(13), 2551–2560 (2008). doi:10.1016/j.dam.2008.11.008
19. Caramia, M., Dell'Olmo, P., Italiano, G.F.: Checkcol: improved local search for graph coloringstar. J. Discret. Algorithms **4**(2), 277–298 (2006). doi:10.1016/j.jda.2005.03.006
20. Mylopoulos, J., Reiter, R.: Order-based genetic algorithms and the graph coloring problem. In: Mylopoulos, J., Reiter, R. (eds.) Handbook of Genetic Algorithms, pp. 72–90. Van Nostrand Reinhold, New York (1991)
21. Fleurent, C., Ferland, J.: Genetic and hybrid algorithms for graph coloring. Ann. Oper. Res. **63**, 437–464 (1996)
22. Plumettaz, M., Schindl, D., Zufferey, N.: Ant local search and its efficient adaptation to graph colouring. J. Oper. Res. Soc. **61**(5), 819–826 (2010). doi:10.1057/jors.2009.27
23. Galinier, P., Hao, J.-K.: Hybrid evolutionary algorithms for graph coloring. J. Comb. Optim. **3**(4), 379–397 (1999). doi:10.1023/A:1009823419804
24. Lü, Z., Hao, J.-K.: A memetic algorithm for graph coloring. Eur. J. Oper. Res. **203**(1), 241–250 (2010). doi:10.1016/j.ejor.2009.07.016
25. Wu, Q., Hao, J.-K.: Coloring large graphs based on independent set extraction. Comput. Oper. Res. **39**(2), 283–290 (2012). doi:10.1016/j.cor.2011.04.002
26. Titiloye, Olawale, Crispin, Alan: Graph coloring with a distributed hybrid quantum annealing algorithm. In: O'Shea, James, Nguyen, Ngoc Thanh, Crockett, Keeley, Howlett, Robert J., Jain, Lakhmi C. (eds.) KES-AMSTA 2011. LNCS, vol. 6682, pp. 553–562. Springer, Heidelberg (2011)
27. Titiloye, O., Crispin, A.: Quantum annealing of the graph coloring problem. Discret. Optim. **8**(2), 376–384 (2011). doi:10.1016/j.disopt.2010.12.001
28. Titiloye, O., Crispin, A.: Parameter tuning patterns for random graph coloring with quantum annealing. PLoS ONE **7**(11), e50060 (2012). doi:10.1371/journal.pone.0050060
29. Moalic, L., Gondran, A.: Variations on Memetic Algorithms for Graph Coloring Problems. http://arxiv.org/abs/arXiv1401.2184
30. Johnson, D.S., Trick, M. (eds.): Cliques, Coloring, and Satisfiability: Second DIMACS Implementation Challenge, 1993. DIMACS Series in Discrete Mathematics and Theoretical Computer Science, vol. 26. American Mathematical Society, Providence (1996)
31. Galinier, P., Hertz, A., Zufferey, N.: An adaptive memory algorithm for the *k*-coloring problem. Discret. Appl. Math. **156**(2), 267–279 (2008). doi:10.1016/j.dam.2006.07.017

The Sim-EA Algorithm with Operator Autoadaptation for the Multiobjective Firefighter Problem

Krzysztof Michalak[✉]

Department of Information Technologies, Institute of Business Informatics,
Wroclaw University of Economics, Wroclaw, Poland
krzysztof.michalak@ue.wroc.pl

Abstract. The firefighter problem is a graph-based optimization problem that can be used for modelling the spread of fires, and also for studying the dynamics of epidemics. Recently, this problem gained interest from the softcomputing research community and papers were published on applications of ant colony optimization and evolutionary algorithms to this problem. Also, the multiobjective version of the problem was formulated.

In this paper a multipopulation algorithm Sim-EA is applied to the multiobjective version of the firefighter problem. The algorithm optimizes firefighter assignment for a predefined set of weight vectors which determine the importance of individual objectives. A migration mechanism is used for improving the effectiveness of the algorithm.

Obtained results confirm that the multipopulation approach works better than the decomposition approach in which a single specimen is assigned to each direction. Given less computational resources than the decomposition approach, the Sim-EA algorithm produces better results than a decomposition-based algorithm.

Keywords: Multipopulation algorithms · Multi-objective evolutionary optimization · Graph-based optimization · Firefighter problem

1 Introduction

The firefighter problem is a graph-based optimization problem formalized in 1995 by Hartnell [13]. This problem can be used for modelling the spread of fires, and also for studying the dynamics of epidemics. The original version of the firefighter problem was single objective. Recently [15], a multiobjective version of the firefighter problem has been proposed.

Developments concerning the firefighter problem can be divided into several areas. Some papers deal with theoretical properties and discuss specific types of graphs and specific problem cases. For example the paper [10] gives lower and upper bounds on the amount of firefighters needed to control a fire in the case of specific planar grids.

© Springer International Publishing Switzerland 2015
G. Ochoa and F. Chicano (Eds.): EvoCOP 2015, LNCS 9026, pp. 184–196, 2015.
DOI: 10.1007/978-3-319-16468-7_16

Some attempts on this problem were also made by the optimization research community. For example, the paper [8] uses a linear integer programming model to solve the single-objective version of the problem. The applications of meta-heuristic methods to the firefighter problem are relatively recent. The paper [3] published in 2014 applies the Ant Colony Optimization (ACO) approach to the single-objective version of the firefighter problem. The authors of this paper have stated, that before its publication not a single metaheuristic approach has been applied to this problem. In the paper [15], published later in 2014 a multiobjective version of the firefighter problem is tackled using the NSGA-II algorithm [6] with an autoadaptation mechanism selecting the best performing genetic operators.

In this paper the multiobjective version of the problem is approached, however, it is treated a little differently than in [15]. Instead of finding the Pareto front we are interested in finding optimal firefighter assignments for a predefined set of N_{sub} weight vectors $W_0 = \left\{\lambda^{(1)}, \ldots, \lambda^{(N_{sub})}\right\}$ which determine the directions of the search as follows. Denote the number of objectives as m. A vector $\lambda^{(j)} = \left[\lambda_1^{(j)}, \ldots, \lambda_m^{(j)}\right]$ determines a search direction in which the first objective has weight $\lambda_1^{(j)}$, the second $\lambda_2^{(j)}$ and so on. For example, a search along direction determined by the vector $\lambda = [1/3, 2/3]$ generates strategies which are best suited if we are interested twice as much in maximizing objective f_2 as in maximizing f_1. This is the approach that is employed in decomposition-based algorithms such as the MOEA/D [14,23]. The approach presented in this paper is intended to be used in research concerning dynamic optimization in the case of the non-deterministic version of the firefighter problem which is currently underway. In the case of dynamic optimization the algorithm may modify the firefighter assignment in response to changes in the environment. This in turn affects the state of the environment. When a decomposition-based approach is used one can simulate the interaction between the algorithm and the environment for each optimization direction separately. Therefore, it is possible to assess in advance, what will happen if certain weights are assigned to the objectives and it is possible to compare various scenarios. To use a Pareto-based algorithm one would have to employ a decision-making module which would choose which strategy should interact with the simulation. Therefore, in this paper a decomposition-based approach is adopted and, instead of the NSGA-II algorithm as in [15], a multipopulation algorithm is used in which subpopulations optimize the solutions along predefined weight vectors from the set W_0. For the same reason the quality of the results is measured by median values of the aggregated objectives rather than using measures typically used for evaluating Pareto fronts such as the hypervolume indicator. The algorithm used in this paper is based on the previous work [16] which introduced the Sim-EA algorithm. This multipopulation algorithm is combined with the operator autoadaptation mechanism used in [15].

To sum up, while this paper uses some of the techniques presented in previous works it introduces the following new elements. Compared to the previous work on the multiobjective firefighter problem [15] the approach was changed from a Pareto-based to a decomposition-based optimization. Following that a different algorithm was employed. With respect to [16] the obvious difference is that this paper tackles a different problem. From this follows that a different measure for choosing subpopulations for migration has to be used. Also, an operator autoadaptation mechanism is used in this paper and an additional migration strategy is proposed.

The rest of the paper is structured as follows. In Sect. 2 the single and multiobjective versions of the firefighter problem are defined. Section 3 presents the algorithm proposed for solving the multiobjective firefighter problem. Section 4 describes the experimental setup and presents the results. Section 5 concludes the paper.

2   Problem Definition

The single objective version of the firefighter problem can be formalized as follows. An undirected graph $G = \langle V, E \rangle$ with N_v vertices is given. Each vertex of the graph G can be labeled using labels from the set L = 'B', 'D', 'U' with the interpretation 'B' = burning, 'D' = defended and 'U' = untouched. Further, we will use a function $l : V \to L$ to denote the labelling of the vertices of the graph G. The spread of fire is simulated in discrete time steps. Initially, vertices from a non-empty set $\emptyset \neq S \subset V$ are labelled 'B' and the remaining ones are labelled 'U'. In each of the following steps of the simulation two events occur. First, a predefined number N_f of still untouched nodes of the graph G (labelled 'U') become defended by firefighters (i.e. become labelled 'D'). Then the fire spreads from the nodes labelled 'B' to all the neighbouring nodes labelled 'U'. The nodes marked 'D' remain defended until the end of the simulation and the fire cannot spread to them. The simulation is stopped when either the fire is contained (i.e. there are no nodes labelled 'U' to which the fire could spread via the edges of the graph G without having to go through a defended node) or when all the undefended nodes are burning. The order in which firefighters are assigned can be represented as a permutation P of the numbers $1, \ldots, N_v$. When firefighters are assigned the first N_f yet untouched nodes are taken from the permutation P. Thus, in every time step exactly N_f firefighters are assigned (with the exception of the final time step in which the number of the untouched nodes may be less than N_f).

In the single objective version of the problem the goal is to find such permutation P that maximizes the number of non-burning nodes at the end of the simulation. In the multiobjective version there are m values $v_i(v)$, $i = 1, \ldots, m$ assigned to each node v in the graph. Each of the v_i values for a given node v represents the value of this node with respect to a certain criterion. In the context of fire containment these criteria could represent, for example, the financial value $v_1(v)$ and the cultural importance $v_2(v)$ of the items stored at the node v.

To calculate the values of the objectives f_i, $i = 1, \ldots, m$ attained by a given permutation P one has to perform the simulation of the fire spread. After the simulation is finished, the values of the objectives are calculated as:

$$f_i = \sum_{v \in V : l(v) \neq' B'} v_i(v) \tag{1}$$

where:

$v_i(v)$ is the value of a given node according to the i-th criterion.

As mentioned in the introduction, the approach adopted in this paper is a decomposition-based optimization. Therefore, we actually search for a set of N_{sub} solutions each of which is the best possible one for a subproblem in which the objective function is an aggregation of the original objectives f_i parameterized by a certain weight vector $\lambda^{(j)} = \left[\lambda_1^{(j)}, \ldots, \lambda_m^{(j)}\right]$, where $j = 1, \ldots, N_{sub}$.

3   The Sim-EA Algorithm

The Sim-EA algorithm presented in this paper is a multipopulation approach. Instead of forming one big population the specimens are divided into N_{sub} sub-populations, each consisting of N_{pop} specimens and a migration mechanism is used to facilitate information transfer between subpopulations as presented in Fig. 1. Thus, the Sim-EA algorithm is based on an island model [21].

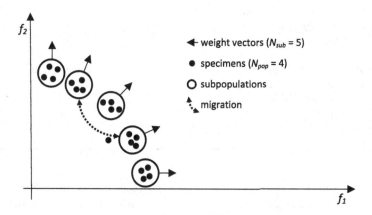

Fig. 1. An overview of elements of the Sim-EA algorithm.

The general outline of the algorithm is based on the previous work [16] in which the Sim-EA algorithm was used for optimizing several similar instances of the TSP problem at once. The overview of the Sim-EA algorithm is presented in Algorithm 1.

3.1   Migration

The key element of the Sim-EA algorithm is the migration mechanism which is based on a function used to determine how likely the specimens are to migrate

Algorithm 1. The overview of the Sim-EA algorithm [16].

IN: N_{gen} - the number of generations
       N_{pop} - the size of each subpopulation
       N_{sub} - the number of subpopulations
       N_{imig} - the number of migrated specimens

Calculate the matrix $S_{[N_{sub} \times N_{sub}]}$ using equation (2)

Initialize subpopulations $P_1, P_2, \ldots, P_{N_{sub}}$.
Evaluate specimens in $P_1, P_2, \ldots, P_{N_{sub}}$.
for $g = 1, \ldots, N_{gen}$ **do**
   Apply genetic operators
   Evaluate newly created specimens
   for $d = 1, \ldots, N_{sub}$ **do**   // Source populations
      $s = \text{SelectSourcePopulation}(S, d)$
      $P'_d =$ the N_{imig} best specimens from P_s
   end for
   for $d = 1, \ldots, N_{sub}$ **do**   // Migration
      for $x \in P'_d$ **do**
         Evaluate x using $\lambda^{(d)}$
         $P'_d = P'_d - \{x\}$
         $w =$ the weakest specimen in P_d
         $P_d = P_d - \{w\}$
         $b = \text{BinaryTournament}(w, x)$
         $P_d = P_d \cup \{b\}$
      end for
   end for
   for $d = 1, \ldots, N_{sub}$ **do**   // Elitist selection
      $e =$ the best specimen in P_d
      $P_d = \text{Select}(P_d \backslash \{e\}, N_{pop} - 1)$
      $P_d = P_d \cup \{e\}$
   end for
end for

---

between two given subpopulations. In the previous work squared differences between cost matrices used in TSP instances were used. In this paper all subpopulations solve the same instance of the firefighter problem (i.e. they work on the same graph G), but have objective functions aggregated using different weight vectors $\lambda^{(j)}$. Consequently, the function $S_{i,j}$ used for determining how likely the specimens are to migrate between subproblems parameterized by weight vectors $\lambda^{(i)}$ and $\lambda^{(j)}$ is calculated as the dot product:

$$S_{i,j} = \lambda^{(i)} \cdot \lambda^{(j)}. \tag{2}$$

In preliminary tests two approaches were considered, one in which the dot product (2) is calculated using normalized vectors and a second one using non-normalized vectors. Since the latter approach yielded better results it was used in the experiments presented in this paper. Note, that the values of the dot product (2) may be affected by the scaling of the objectives. In this paper both objectives were of the same magnitude, but in general it might be necessary to normalize the objectives before calculating the dot product (2).

The migration is performed in two phases. First, for each subpopulation P_d, $d = 1, \ldots, N_{sub}$ a source population P_s is determined according to the migration strategy. A set of immigrants P'_d is built by selecting the N_{imig} best specimens from the population P_s. In the second phase the immigrants from the sets P'_d are merged into the populations P_d for $d = 1, \ldots, N_{sub}$. During the merge phase each immigrant participates in a binary tournament [17] against the currently weakest specimen in the existing population P_d. If the immigrant has higher value of the objective function aggregated using $\lambda^{(d)}$ it replaces the weakest specimen in the population P_d.

The selection of the source subpopulations P_s depends on the adopted migration strategy. In the **"rank"** strategy all subpopulations $P_{s'}$ except P_d are ranked according to the increasing values of the dot product (2). From them one subpopulation P_s is selected using roulette wheel selection with probabilities proportional to the ranks of the subpopulations. In the **"highest"** strategy the source population P_s is the one with the highest value of the dot product (2), **"uniform"** is a strategy in which the source population P_s is selected randomly with uniform probability distribution among all populations except P_d and **"none"** means that no migration is performed.

3.2   Operator Autoadaptation

In the previous paper [15] in which the multiobjective firefighter problem was tackled using the NSGA-II algorithm an operator autoadaptation mechanism was used. From the results presented in the aforementioned paper it follows that various operators (e.g. different crossover procedures) perform well on various stages of the search. Therefore, in this paper an autoadaptation mechanism was also used. The autoadaptation mechanism used in this paper is based on success rates of the genetic operators. It counts the number of times each operator was used n_i and the number of improvements obtained b_i. Since in this paper a decomposition-based approach is used the improvements are determined with respect to the aggregated objective function not with respect to the individual objectives. However, after one application of the crossover operator the value of b_i can increase by more than 1, because each offspring improving on each parent is counted separately. Thus, a maximum increase of 4 can be obtained (both offspring improving on both parents). From the n_i and b_i values the success rate $s_i = b_i/n_i$ is calculated (if $n_i = 0$ then $s_i = 0$).

The s_i values are used to calculate probabilities with which individual operators are used. A minimum probability P_{min} is given to each of the N_{op} operators and the remaining $1 - N_{op}P_{min}$ is divided proportionally to success rates s_i of the operators. When an operator is to be applied a roulette wheel selection is performed. The selection of crossover and mutation operators is performed separately.

In this paper a set of 10 crossover operators and 5 mutation operators was used for genetic operations. The crossover operators were: Cycle Crossover (CX) [19], Linear Order Crossover (LOX) [9], Merging Crossover (MOX) [18], Non-Wrapping Order Crossover (NWOX) [5], Order Based Crossover (OBX) [20], Order Crossover (OX) [11], Position Based Crossover (PBX) [20], Partially Mapped Crossover (PMX) [12], Precedence Preservative Crossover (PPX) [1,2] and Uniform Partially Mapped Crossover (UPMX) [4]. The mutation operators were: displacement mutation, insertion mutation, inversion mutation, scramble mutation and transpose mutation.

Due to limited space in this paper it is not possible to present details concerning the performance of the operators, but in general it can be stated that the CX and PPX crossover operators performed best, while the OX crossover operator performed worst. Among the mutation operators the displacement mutation performed best with scramble and inversion mutation only a little worse. The transpose mutation performed worst.

4   Experiments and Results

In the experiments the Sim-EA algorithm with various migration strategies was tested. Also, a single population approach known from the MOEA/D algorithm was used for comparison. The MOEA/D algorithm used the same operators and the autoadaptation mechanism as the Sim-EA.

Test data sets were prepared as follows for graph size $N_v = 50, 75, 100, 125, 150, 175, 200, 225$ and 250. The graph G was generated by randomly determining, for each pair of vertices v_i, v_j, if there exists an edge $\langle v_i, v_j \rangle$. The probability of generating an edge was set to $P_{edge} = 2.5/N_v$ in order to ensure that the average number of edges adjacent to a vertex was similar for all the instances. Costs were assigned to all vertices of the graph G by drawing pairs of random values with uniform probability on a triangle formed by points $[0, 0]$, $[0, 100]$, $[100, 0]$. Such an assignment ensures that it is not possible to maximize both objectives at the same time, because the sum of costs associated with a vertex cannot exceed 100.

In the experiments the performance of the Sim-EA algorithm with four migration strategies (highest, rank, uniform and none) and the MOEA/D algorithm with three different parameter settings was tested. The number of subproblems was set to $N_{sub} = 20$.

The Sim-EA algorithm was parameterized by setting the size of each of the subpopulations to the instance size N_v. Therefore, the total number of specimens used by the Sim-EA algorithm was $N_{spec} = 20 \cdot N_v$. The number of generations was set to $N_{gen} = 250$ generations for all problem instances.

The MOEA/D algorithm does not use subpopulations, therefore the parameter setting had to be different. The size of the population is determined by a parameter H - the size of a step used for generating weight vectors. Because there is exactly one weight vector assigned to each specimen the population size is equal to the number of weight vectors. For an m-objective problem, weight vectors $\lambda^{(j)}$ are generated by selecting all the possible m-element combinations of numbers from the set:

$$\left\{ \frac{0}{H}, \frac{1}{H}, \ldots, \frac{H}{H} \right\} \tag{3}$$

such that:

$$\sum_{i=1}^{m} \lambda_i^{(j)} = 1. \tag{4}$$

The number of weight vectors and at the same time the number of specimens is $N_{spec} = C_{H+m-1}^{m-1}$. For a biobjective optimization problem ($m = 2$) the set of weight vectors is $W = \{\lambda^{(j)}\}$, $j \in \{0, \ldots, H\}$, where:

$$\lambda^{(j)} = \left[\frac{j}{H}, \frac{H-j}{H} \right]. \tag{5}$$

Because in this paper we are interested in finding optimal solutions along $N_{sub} = 20$ predefined directions the number of specimens in the experiments had to be chosen in such a way, that each of the N_{sub} weight vectors is present in the set W (i.e. $W_0 \subset W$). In cases when the population size N_{spec} was larger than N_{sub} additional N_i weight vectors were added between the main N_{sub} predefined weight vectors as presented in Fig. 2.

Three different sets of parameters were used with the MOEA/D algorithm (cf. Table 1). In the first parametrization $MOEA/D_{20}$ the number of specimens was 20 for all data sets (i.e. $N_{spec} = N_{sub}$). In the two remaining parameter sets a larger population was used. From this larger population $N_{sub} = 20$ specimens were associated with weight vectors from the set W_0 which represented the main search directions (i.e. the aggregations for which the best solutions had

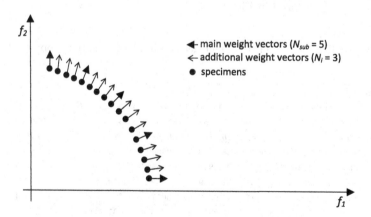

Fig. 2. The configuration of weight vectors in the MOEA/D algorithm used for optimization along $N_{sub} = 5$ main directions with the total population size $N_{spec} = 17$.

Table 1. Parameter sets used in the experiments with MOEA/D algorithm.

N_v	$MOEA/D_{20}$			$MOEA/D_{210}$			$MOEA/D_{N_{spec}}$		
	N_{spec}	N_{gen}	N_i	N_{spec}	N_{gen}	N_i	N_{spec}	N_{gen}	N_i
50		400			200		1008	50	52
75		1000			500		1502	70	78
100		2000			750		2015	80	105
125		3000			1000		2509	90	131
150	20	5000	0	210	1400	10	3003	100	157
175		7000			1800		3516	110	184
200		10000			2700		4010	150	210
225		12000			2800		4504	160	236
250		15000			3600		5017	200	263

to be found). The remaining specimens were associated with search directions between the main N_{sub} vectors. In the $MOEA/D_{210}$ parametrization $N_i = 10$ specimens were added between each two main search directions, thus the total population size was 210. In the $MOEA/D_{N_{spec}}$ parameter set the total number of specimens in the MOEA/D algorithm was set to a number a little larger than the total number of specimens N_{spec} used in the Sim-EA algorithm. To achieve this a varying number N_i of additional specimens were used. To make a fair comparison, the number of generations for each MOEA/D parameter set was adjusted in such a way that the minimum runtime of the MOEA/D in each repetition of the test was slightly larger than the maximum runtime of the Sim-EA algorithm.

For each problem instance 30 repetitions of the test were performed for each of the four migration strategies and for each parameter set used with the MOEA/D algorithm. After completing N_{gen} generations the best result obtained along each of the N_{sub} directions was recorded. From the gathered results median values obtained in the 30 iterations were calculated. The results for all test instances are summarized in Table 2. For each data set the best result is underlined. The results are also presented in Fig. 3.

Table 2. Median values of the aggregated objective functions obtained in the experiments.

N_v	Sim-EA				MOEA/D		
	None	Highest	Rank	Uniform	$MOEA/D_{20}$	$MOEA/D_{210}$	$MOEA/D_{N_{spec}}$
50	541.6037	566.8921	621.2476	620.8365	486.5366	532.4463	536.6768
75	642.7674	653.4373	714.5162	720.6106	545.0088	587.8820	590.2881
100	673.7599	687.6201	742.6952	738.5506	557.9805	582.8894	587.8565
125	736.4336	753.2498	792.7041	790.5849	739.4365	763.5474	780.1217
150	772.5983	787.7714	857.3251	851.2659	612.6652	640.4161	638.2581
175	783.4265	803.4243	838.2117	835.0432	636.6017	653.1362	657.8071
200	773.0628	788.0596	854.1746	848.2873	614.7756	637.8039	642.2385
225	876.5625	892.6258	948.5553	934.2622	651.9642	663.5777	693.1615
250	905.2262	925.0562	962.4042	971.3175	775.4714	794.873	840.8401

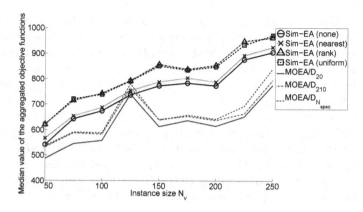

Fig. 3. Median values of the aggregated objective functions obtained in the experiments.

From the results presented in Table 2 it is clear that the Sim-EA algorithm produces the best results mainly with the "rank" migration strategy. The "uniform" migration strategy usually gives a little worse results, but for two datasets it outperformed the "rank" strategy. Graphs presented in Fig. 3 show that the median value of the objectives obtained by the Sim-EA algorithm increases approximately linearly with the instance size N_v. In the case of the MOEA/D algorithm there is a sharp increase in the obtained median values for $N_v = 125$ and a decrease for the range $N_v \in [150, 200]$.

In order to verify the statistical significance of the results the Wilcoxon rank test [22] was performed. This particular test was used because it does not assume the normality of the distributions which may be hard to ensure in practical applications. Also, this test was recommended in a recent survey [7] as one of the methods suitable for statistical comparison of results produced by metaheuristic optimization algorithms. Table 3 presents, for each instance size N_v, the best and

Table 3. Best performing algorithms and the results of the statistical verification.

N_v	Best (1st)	Second-best (2nd)	p-value $1^{st} : 2^{nd}$	p-value rank : none
50	Sim-EA (rank)	Sim-EA (uniform)	$4.0938 \cdot 10^{-1}$	$2.5220 \cdot 10^{-89}$
75	Sim-EA (uniform)	Sim-EA (rank)	$3.5698 \cdot 10^{-1}$	$6.1642 \cdot 10^{-93}$
100	Sim-EA (rank)	Sim-EA (uniform)	$2.5766 \cdot 10^{-6}$	$1.8897 \cdot 10^{-90}$
125	Sim-EA (rank)	Sim-EA (uniform)	$2.3406 \cdot 10^{-7}$	$3.6572 \cdot 10^{-93}$
150	Sim-EA (rank)	Sim-EA (uniform)	$8.1925 \cdot 10^{-14}$	$1.5537 \cdot 10^{-94}$
175	Sim-EA (rank)	Sim-EA (uniform)	$3.9639 \cdot 10^{-1}$	$2.2641 \cdot 10^{-88}$
200	Sim-EA (rank)	Sim-EA (uniform)	$1.4020 \cdot 10^{-2}$	$1.4035 \cdot 10^{-94}$
225	Sim-EA (rank)	Sim-EA (uniform)	$1.9732 \cdot 10^{-5}$	$1.1567 \cdot 10^{-89}$
250	Sim-EA (uniform)	Sim-EA (rank)	$1.4062 \cdot 10^{-2}$	$1.7508 \cdot 10^{-88}$

the second-best algorithm. The "p-value $1^{st} : 2^{nd}$" is the result of a statistical test with the null hypothesis that the median values are the same for both algorithms. The "p-value rank : none" is the result of a statistical test comparing the Sim-EA algorithm with rank-based migration and the Sim-EA algorithm without migration. The p-values that correspond to the significance of 95 % or better (i.e. p-value ≤ 0.05) are underlined.

Comparison of the best and the second-best algorithm for $N_v = 100$, 125, 150, 200 and 225 confirmed that the difference in favour of the "rank" strategy is significant at the significance level 0.05. In the case of $N_v = 50$ and 175 the significance was not confirmed. Obviously, for $N_v = 75$ and 250 the differences are in favour of the "uniform" strategy. These differences were shown to be insignificant in the case of $N_v = 75$ and significant in the case of $N_v = 250$.

5   Conclusion

In this paper the multipopulation algorithm Sim-EA was applied to the multi-objective firefighter problem. In the experiments it was observed that the multipopulation approach can utilize computational resources more effectively than a traditional decomposition-based method. The best performing migration scheme is the "rank" scheme in which the source population for migration is randomly chosen with probability proportional to the rank based on the dot product of weight vectors associated with the source and destination subpopulations. A little worse performing migration scheme is the "uniform" scheme in which the source population is randomly chosen with uniformly distributed probability. For two datasets, however, it outperformed the "rank" scheme. Migration significantly improves the results obtained by the Sim-EA algorithm. Statistical comparison of the Sim-EA with the rank-based migration with Sim-EA without migration confirmed with a very high statistical significance that the rank-based Sim-EA performs better.

Further work on the presented topic is currently underway and it is focused on applying the approach described in this paper to dynamic optimization of a non-deterministic version of the firefighter problem.

References

1. Bierwirth, C., Mattfeld, D.C., Kopfer, H.: On permutation representations for scheduling problems. In: Proceedings of the 4th International Conference on Parallel Problem Solving from Nature, pp. 310–318. Springer (1996)
2. Blanton, J.L., Jr., Wainwright, R.L.: Multiple vehicle routing with time and capacity constraints using genetic algorithms. In: Proceedings of the 5th International Conference on Genetic Algorithms, pp. 452–459. Morgan Kaufmann Publishers Inc., San Francisco, CA, USA (1993)
3. Blum, C., Blesa, M.J., García-Martínez, C., Rodríguez, F.J., Lozano, M.: The firefighter problem: application of hybrid ant colony optimization algorithms. In: Blum, C., Ochoa, G. (eds.) EvoCOP 2014. LNCS, vol. 8600, pp. 218–229. Springer, Heidelberg (2014)

4. Cicirello, V.A., Smith, S.F.: Modeling GA performance for control parameter optimization. In: Whitley, L. (ed.) GECCO-2000: Proceedings of the Genetic and Evolutionary Computation Conference: A Joint Meeting of the Ninth International Conference on Genetic Algorithms (ICGA-2000) and the Fifth Annual Genetic Programming Conference (GP-2000). Morgan Kaufmann Publishers, Massachusetts (2000)

5. Cicirello, V.A.: Non-wrapping order crossover: an order preserving crossover operator that respects absolute position. In: Proceedings of the 8th Annual Conference on Genetic and Evolutionary Computation, pp. 1125–1132. ACM, New York, NY, USA (2006)

6. Deb, K., Pratap, A., Agarwal, S., Meyarivan, T.: A fast and elitist multiobjective genetic algorithm: NSGA-II. IEEE Trans. Evolut. Comput. **6**, 182–197 (2002)

7. Derrac, J., Garca, S., Molina, D., Herrera, F.: A practical tutorial on the use of nonparametric statistical tests as a methodology for comparing evolutionary and swarm intelligence algorithms. Swarm Evol. Comput. **1**(1), 3–18 (2011)

8. Develin, M., Hartke, S.G.: Fire containment in grids of dimension three and higher. Discrete Appl. Math. **155**(17), 2257–2268 (2007)

9. Falkenauer, E., Bouffouix, S.: A genetic algorithm for job shop. In: Proceedings of the 1991 IEEE International Conference on Robotics and Automation, pp. 824–829 (1991)

10. Feldheim, O.N., Hod, R.: 3/2 firefighters are not enough. Discre. Appl. Math. **161**(12), 301–306 (2013)

11. Goldberg, D.: Genetic Algorithms in Search, Optimization, and Machine Learning. Addison Wesley, Reading (1989)

12. Goldberg, D.E., Lingle Jr, R.: Alleles, loci, and the traveling salesman problem. In: Grefenstette, J.J. (ed.) Proceedings of the First International Conference on Genetic Algorithms and Their Applications, pp. 154–159. Lawrence Erlbaum Associates Publishers, Hillsdale (1985)

13. Hartnell, B.: Firefighter! an application of domination. In: 20th Conference on Numerical Mathematics and Computing (1995)

14. Li, H., Zhang, Q.: Multiobjective optimization problems with complicated pareto sets, MOEA/D and NSGA-II. IEEE Trans. Evolut. Comput. **13**(2), 284–302 (2009)

15. Michalak, K.: Auto-adaptation of genetic operators for multi-objective optimization in the firefighter problem. In: Corchado, E., Lozano, J.A., Quintián, H., Yin, H. (eds.) IDEAL 2014. LNCS, vol. 8669, pp. 484–491. Springer, Heidelberg (2014)

16. Michalak, K.: Sim-EA: an evolutionary algorithm based on problem similarity. In: Corchado, E., Lozano, J.A., Quintián, H., Yin, H. (eds.) IDEAL 2014. LNCS, vol. 8669, pp. 191–198. Springer, Heidelberg (2014)

17. Miller, B.L., Goldberg, D.E.: Genetic algorithms, tournament selection, and the effects of noise. Complex Syst. **9**, 193–212 (1995)

18. Mumford, C.L.: New order-based crossovers for the graph coloring problem. In: Runarsson, T.P., Beyer, H.G., Burke, E., Merelo-Guervós, J.J., Darrell Whitley, L., Yao, X. (eds.) PPSN IX. LNCS, vol. 4193, pp. 880–889. Springer, Heidelberg (2006)

19. Oliver, I.M., Smith, D.J., Holland, J.R.C.: A study of permutation crossover operators on the traveling salesman problem. In: Proceedings of the Second International Conference on Genetic Algorithms on Genetic Algorithms and Their Applications, pp. 224–230. Lawrence Erlbaum Associates Inc., Hillsdale, NJ, USA (1987)

20. Syswerda, G.: Schedule optimization using genetic algorithms. In: Davis, L. (ed.) Handbook of Genetic Algorithms. Van Nostrand Reinhold, New York (1991)

21. Whitley, D., Rana, S., Heckendorn, R.B.: The island model genetic algorithm: on separability, population size and convergence. J. Comput. Inf. Technol. **7**, 33–47 (1998)
22. Wilcoxon, F.: Individual comparisons by ranking methods. Biometrics Bull. **1**(6), 80–83 (1945)
23. Zhang, Q., Li, H.: MOEA/D: A multiobjective evolutionary algorithm based on decomposition. IEEE Trans. Evolut. Comput. **11**(6), 712–731 (2007)

True Pareto Fronts for Multi-objective AI Planning Instances

Alexandre Quemy$^{(\boxtimes)}$ and Marc Schoenauer

TAO Project, INRIA Saclay and LRI - University of Paris-Sud and CNRS,
Orsay, France
{alexandre.quemy,marc.schoenauer}@inria.fr

Abstract. Multi-objective AI planning suffers from a lack of benchmarks with known Pareto Fronts. A tunable benchmark generator is proposed, together with a specific solver that provably computes the true
Pareto Front of the resulting instances. A wide range of Pareto Front
shapes of various difficulty can be obtained by varying the parameters
of the generator. The experimental performances of an actual implementation of the exact solver are demonstrated, and some large instances
with remarkable Pareto Front shapes are proposed, that will hopefully
become standard benchmarks of the AI planning domain.

1 Introduction

Contrary to single objective problems, Multi-Objective Problems (MOP) involve
several contradictory criteria to be optimized. This distinction entails a modification of the concept of optimality itself: the optimal solution of a MOP is not
a single solution but a set of solutions that represents trade-offs known as the
Pareto Set. This set is made of the non-dominated points of the search space, i.e.
the solutions that cannot be improved w.r.t. one objective without deteriorate
at least another one. Formally, x dominates y if $\forall i \in \{1, \ldots, n\}$, $f_i(x) \succeq f_i(y)$
and $\exists j \in \{1, \ldots, n\}$, $f_j(x) \succ f_j(y)$. The projection of the Pareto Set over the
objective space is called the Pareto Front.

Many benchmark suites exist for continuous multi-objective optimization
(the famous ZDT [9], IHR [1], . . .), for which the exact Pareto Front can be
analytically computed, and with known difficulties (e.g. dimensionality, shape
of the Pareto Fronts, existence of local Pareto-optima, . . .). For combinatorial
optimization, however, the situation is not yet so clear, and whereas there exist
famous benchmark problems of all sizes, their true Pareto Fronts are only exactly
known for the simplest problems (see e.g., MOCOLIB[1], offering several instances
of several well-known combinatorial benchmark problems).

The benchmark suite introduced in the present work is concerned with *AI
planning*: A planning domain D is defined by a set of predicates that define

---

[1] http://www.mcdmsociety.org/MCDMlib.html.

© Springer International Publishing Switzerland 2015
G. Ochoa and F. Chicano (Eds.): EvoCOP 2015, LNCS 9026, pp. 197–208, 2015.
DOI: 10.1007/978-3-319-16468-7_17

the state of the system when instantiated and a set of possible actions that can be triggered in states where their pre-conditions are satisfied, resulting in a new state. A planning problem instance $\mathcal{P}_D(I, G)$ is defined on a given planning domain D by a list of objects, used to instantiate the predicates to define the states, an initial state I and a goal state G. The aim is to come up with an optimal *feasible plan*, i.e., a set of actions that, when applied in turn to the initial state, lead the system to the goal state, and is optimal w.r.t. a given measure: the number of actions, or the total cost of the plan when actions have non-uniform costs, or the total *makespan* (total duration of the plan) when actions have durations, and can be run in parallel.

MiniZenoTravel is a simple temporal planning domain related to logistics, inspired by the well-known ZenoTravel problem introduced in the 3rd edition of the IPC series[2]. It involves cities, passengers, and planes (see e.g., Fig. 1); Planes can fly from one city to another when a link exists (on Fig. 1, the flight duration is attached to the link); Planes fly either empty, or carrying a unique passenger – and these are the only possible actions. A MiniZenoTravel instance is defined by the number of cities and the graph of the possible flights between them, a number of passengers and a number of planes. In the initial state I, all passengers and planes are in city c_I, and in the goal state G, all passengers must be in city c_G. Previous work proposed a multi-objective version of these benchmarks called MultiZenoTravel, by adding a *cost* for landing in some cities: the second objective is to minimize the *total cost* of the plan [2,8]. The latter work demonstrated that such problems could provide Pareto Fronts of various shapes and difficulties. However, the authors were only able to provide the exact Pareto Front for very small instances, due to the combinatorial explosion of the solution space.

The present work formally analyzes the MultiZenoTravel benchmarks and provides an algorithm to compute their true Pareto fronts in reasonable time, even for very large instances. Beyond providing a generic way to generate Pareto Fronts of tunable complexities for AI Planning, the proposed MultiZeno-Travel benchmarks will allow testing different multi-objective optimization algorithms, from generic decomposition methods (weighted sum aggregation, Tchebycheff decomposition, Boundary Intersection approach – see e.g., [6]) to Pareto-based Evolutionary Algorithms, on complex benchmarks for which the Pareto Front is exactly known.

The paper is organized as follows: Sect. 2 formally presents the MultiZeno-Travel benchmark, proving some properties of their Pareto optimal plans. Building on these properties, Sect. 3 proposes the ZenoSolver algorithm to actually derive the true Pareto Front for these instances. Sample experimental results demonstrate the diversity of Pareto Fronts that can be obtained, and gives performance measurements of its complexity on large instances.

---

[2] http://ipc.icaps-conference.org/.

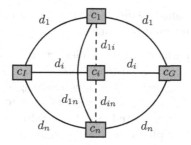

Fig. 1. A schematic view of a general MULTIZENOTRAVEL problem.

2 MULTIZENOTRAVEL Problem

2.1 Instances

Let us introduce some notations related to the planning problem briefly presented in the introduction: a MULTIZENOTRAVEL instance (Fig. 1) is defined by the following elements:

- n central cities, organized as a clique in which every node is connected to C_I and C_G, respectively the initial city and the goal city.
- $c \in (\mathbb{R}^+)^n$, where c_i is the cost for landing in C_i.
- $D \in (\mathbb{R}^+)^{n \times n}$, where D_{ij} is the flying time between C_i and C_j.
- $d^I \in (\mathbb{R}^+)^n$, where d_i^I is the flying time between C_I and C_i.
- $d^G \in (\mathbb{R}^+)^n$, where d_i^G is the flying time between C_i and C_G.
- p planes, initially in C_I, that have a capacity of an unique person.
- t persons, initially in C_I.

As said, the goal is to carry all t persons, initially in c_I, to c_G using p planes, minimizing both the makespan and the cost of the plan. In order to ease the identification of the true Pareto Front, a symmetry constraint is added: $\forall i \in [1, n], d_i^I = d_i^G$ and from thereon we will refer to a unique vector d.

Without loss of generality, all pairs (d_i, c_i) are assumed to be pairwise distinct. Otherwise, the 2 cities can be "merged" and the resulting $n - 1$ cities problem is equivalent to the original n cities problem, as there exist no city capacity constraints. Finally, we only consider cases where $t \geq p$, as the problem is otherwise trivial.

2.2 Pareto Optimal Plans

Let us make another simplifying assumption:
Assumption A1: $\forall (i, j) \in [1, n]^2, d_i + d_j < d_{ij}$[3]. Then the following holds.

---

[3] This might look unrealistic in real-world logistic domain. However, we hypothesize that the proposition still holds with the weaker condition that for any cities C_i, C_j, C_k (if we state the cost of C_I and C_G are respectively C_0 and C_{n+1}), $d_{ik} \leq d_{ij} + d_{jk}$ (triangle inequality).

Proposition: Pareto-optimal plans are plans where exactly $2t - p$ (possibly identical) central cities are used by a flight.

Proof: Consider a plan where a person flies from C_i to C_j. Using the same plane, the same person could fly instead from C_i to C_G, and the plane would return empty to C_j. The plan could continue unchanged from thereon: because of the hypothesis on makespans, the needed resource would be in C_j on time. Moreover, the total cost is unchanged, and the total makespan is lower or equal to the original one: the new plan thus Pareto-dominates the original one.

Iterating the same reasoning for each person, and each empty plane, we conclude that there are no flights between central cities in Pareto-optimal plans. Thus bringing the t persons from C_I to C_G will amount to carry each person through one central city: t flights will be needed from C_I to one C_i, then t flights from C_i to C_G. Finally, because planes do not need to come back from C_G in the end, only $t - p$ flights back empty will be needed, possibly through some different central cities – hence the result.   □

PPPs and Admissible PPPs: According to the above proposition, a Possibly Pareto-optimal Plan (PPP) is defined by 2 tuples, namely $e \in [0, n]^t$ for cities involved in eastbound flights, and $w \in [0, n]^{t-p}$ for westbound flights. Nevertheless, e and w do not hold any information about which plane will land in a particular city. This is the reason why there exists many feasible schedules, i.e., schedules that actually are feasible plans for p planes[4] using the corresponding $4t - 2p$ edges. There are at most $n^{(2t-p)}$ possible PPP but it is clear that the set of PPPs contains many redundancies, that can easily be removed by ordering the indices:

Definition: An *admissible PPP* is a pair of $E \times W$, where $E = \{e \in [1, n]^t; \forall i \in [1, t-1], d_{e_i} \geq d_{e_{i+1}}\}$ and $W = \{w \in [1, n]^{t-p}; \forall i \in [1, t-p-1], d_{w_i} \geq d_{w_{i+1}}\}$.

Number of admissible PPPs: Let K_k^m be the set of k-multicombinations (or multi-subset of size k) with elements in a set of size m. The cardinality of K_k^m is $\Gamma_k^m = \binom{m+k-1}{k}$. As E is in bijection with K_t^n, and W with K_{t-p}^n, the number of PPP is $\Gamma_t^n \Gamma_{t-p}^n$, i.e., $\binom{n+t-1}{t}\binom{n+(t-p)-1}{t-p}$.

Cost of a PPP: Given the PPP $C = (e, w) \in E \times W$, the cost of **any** plan using only the cities in e and w is uniquely defined by $\text{Cost}(C) = \sum_{e_i \in e} c_{e_i} + \sum_{w_i \in w} c_{w_i}$.

Makespan of a PPP: The makespan of a PPP is thus that of the shortest schedule that uses its $4t - 2p$ edges in a feasible way. Trivial upper and lower bounds for the shortest makespan of a PPP C are respectively $M_S(C)$, the makespan of the sequential plan (i.e., that of the plan for a single plane that would carry all persons one by one), and $M_L(C)$, the makespan of the perfect plan where none of the p planes would ever stay idle. As discussed in Sect. 3, these bounds are useful to prune the set of PPPs.

---

[4] Most of them are probably not Pareto-optimal, but w.r.t the previous proposition, any schedule resulting from a larger tuple e or w would be Pareto-dominated.

$$M_S(C) = 2(\sum_{e_i \in e} d_{e_i} + \sum_{w_i \in w} d_{w_i}) \qquad\qquad M_L(C) = \frac{M_S(C)}{p}$$

Greedy domination: Given two PPP C and C', C *greedily dominates* C' if $M_S(C) \leq M_L(C')$ and $Cost(C) \leq Cost(C')$.

2.3   Computing the Shortest Makespan

Flight Patterns. Clearly, within a PPP, all possible plane moves can be categorized into only 3 patterns:

P1: plane leaves C_I (non empty), flies eastward to city C_i, and goes on to C_G.
P2: plane leaves C_G (empty), flies westward to city C_i, and goes on to C_G.
P3: two planes are involved here; first plane leaves C_I (with a passenger), flies to city C_i, and goes back empty to C_I; second plane leaves C_G empty, flies to C_i, and flies back with the passenger to C_G. Note that there can be some delay between the drop-off of the person at the central city, and the arrival of the second plane.

Given a feasible plan using only the three above patterns, let α_E, α_W, and β be the numbers of effective P1, P2, and P3 patterns respectively. It is clear that β entirely determines α_E and α_W, as $\alpha_W = t - p - \beta$ and $\alpha_E = t - \beta$. Considering a PPP C, it is possible for a given β to have multiple choices for the cities involved in P3. Each choice is denoted β_{set} and the set of β_{set} the β-PowerSet.

The optimal makespan for a given admissible PPP C is the lowest makespan obtained for all $\beta_{\text{set}} \in \beta$-PowerSet. Once the optimal makespan for a couple (C, β_{set}) determined, iterating over the β-PowerSet held by C returns the optimal makespan for C. Finally, iterating the process over the set of PPP returns the Pareto Front for the considered instance.

The method to compute the optimal makespan for a particular couple (C, β_{set}) is broken down into two steps. In a first step, each β_{set} defines a subproblem without any P3 that is easy to solve. The second step is to take into account the P3 patterns in C. After detailing these two steps, we will give a constructive proof that the obtained makespan is optimal.

Step 1: Handling P3-Free PPPs. For a given $((e, w), \beta_{\text{set}})$ denote $e' = e \backslash \beta_{\text{set}}$, i.e. the tuple e from which all elements of β_{set} have been removed, and $w' = w \backslash \beta_{\text{set}}$ defined similarly. As a result, $((e', w'), \emptyset)$ is the subproblem of $((e, w), \beta_{\text{set}})$ that does not contain any P3 ($\beta' = 0$).

For a PPP with $\beta = 0$, greedy Algorithm 1 dispatches the longest flight durations first, assigning them to the available planes with shortest 'private' makespan (who have yet flown the less), ending with the one-way last flights from C_I to C_G (planes end in C_G). The algorithm returns the flight durations (D_k^1) for all planes k (to be used in the second step), and the optimal makespan for the subproblem is obviously $\max_k(D_k^1)$.

Algorithm 1. Computing the optimal makespan of PPP (e, w) when $\beta = 0$

$i \leftarrow 1 ; j \leftarrow 1$ {Indices of cities in e and w resp., longest durations first}

$D_k \leftarrow 0, \; k = 1, \dots, p$ {'Private' makespan for plane k}

$S_k \leftarrow EAST, \; k = 1, \dots, p$ {All planes are in C_I, going eastward}

while $j \leq t - p$ **do**

     $k \leftarrow \text{ArgMin}_i (D_i)$ {Plane with shortest private makespan, in C_I or C_G}

     if $S_k = EAST$ **then**

         $D_k \leftarrow D_k + 2d_{e_i}$ {From C_I to C_G through city C_{e_i}}

         $S_k \leftarrow WEST \; ; \; i \leftarrow i + 1$

     else

         $D_k \leftarrow D_k + 2d_{w_j}$ {From C_G to C_I through city C_{w_j}}

         $S_k \leftarrow EAST \; ; \; j \leftarrow j + 1$

     end if

end while   {Are there persons and planes left in C_I?}

while $i \leq t$ **do**

     $k \leftarrow \underset{i; S_i = EAST}{\text{ArgMin}} (D_i)$ {Plane in C_I with shortest private makespan}

     $D_k \leftarrow D_k + 2d_{e_i}$ {From C_I to C_G through city C_{e_i}}

     $S_k \leftarrow WEST \; ; \; i \leftarrow i + 1$

end while

return $(D_k)_{k=1,\dots,p}$ {All private makespans are needed for the second step}

Makespan $(e, w) = \underset{k=1,\dots,p}{\max} \{D_k\}$

Step 2: Tackling Patterns P3. The second step consists in dispatching the durations of P3 patterns among the planes according to their previous flight durations $(D_k^1)_{k=1,\dots,p}$, by sequentially assigning the longest P3 flight to the two planes with the smallest current flight durations. This can be performed greedily again, with a slightly modified version of the Algorithm 1, if we only consider the flight durations.

However, within a P3 pattern, if the plane coming from C_G lands in the central city before the person has yet arrived from C_I, it has to wait. Consequently, it is possible that the makespan of the plan is not simply the sum of the pattern durations. Indeed, the described algorithm is not taking into account the possibility of a waiting point and this is the reason why we first have to discuss the construction of a feasible plan according to the final vector of durations $(D_k^2)_{k=1,\dots,p}$ before discussing the optimality of $\underset{k=1,\dots,p}{\max} \{D_k^2\}$ as the makespan of the associated PPP.

Proposition: It is always possible to construct a feasible plan with the makespan returned by the two-steps method described above.

Proof by construction: Considering a P3 pattern performed by planes p_1 in C_G and p_2 in C_I through city C_i. Their schedules will look something like:

$$p_1 : C_I \rightarrow \dots \rightarrow C_G \rightarrow \overset{\blacklozenge}{C_i} \rightarrow C_G \rightarrow \dots \rightarrow C_G$$

$$p_2 : C_I \rightarrow \dots \rightarrow C_I \rightarrow \overset{\lozenge}{C_i} \rightarrow C_I \rightarrow \dots \rightarrow C_G$$

Let $\blacklozenge$ denote the time t_1 when plane p_1 arrives in C_i and $\lozenge$ the time t_2 when plane p_2 (with the person) arrives in C_i. If $t_2 > t_1$ then plane p_1 will have to wait $t_2 - t_1$ in C_i before flying back to C_G with the person. But the duration vector (D_k^2) returned by the two steps algorithm is computed assuming no waiting point. Consequently, the proposition is equivalent to assert that we can always build a plan without any waiting point.

In order to do so, for each P3, the idea is to perform the westward part as early as possible, and on the opposite, to perform the eastward part as late as possible, thus ensuring that there is no waiting time.

In order to construct such an optimal plan, we will remember the cities of every plane and every pattern during both previous steps of the algorithm. From there on, let us consider now only planes that have to perform at least one P3 pattern.

1. For each plane, select the one with the maximum number of P3 patterns to be performed. In case of tie, select the plane with longest P3 duration, or the plane with the largest current 'private' makespan.
2. Construct a partial schedule with only P1 and P2 patterns (Step 1 above).
3. For every 'not already started' P3 pattern, add its eastward part at the end of the schedule by descending order of durations.
4. For every 'already started' P3 pattern, add its westward part at the beginning of the schedule by ascending order of durations.   □

Example: Considering $t = 7$, $p = 3$, $d = (2, 4, 6)$, $C = (3, 3, 2, 2, 2, 1, 1)(3, 3, 2, 1)$ and $\beta_{\text{set}} = \{3, 2, 1\}$ leads to the sub-problem $C' = (3, 2, 2, 1)(3)$ with $t' = 4$. Step 1 above gives the 'private' makespans D_k^1 in the table below. Adding the P3 patterns (Step 2) gives the 'private' makespans D_k^2. The complete schedule can then be built according to the method described above.

p_i	D_i^1	D_i^2	P3
p_1	12	32	2
p_2	24	28	1
p_3	8	32	3

$p_1 : c_I \to c_2 \to c_G \to \overset{\blacklozenge_3}{c_3} \to c_G \to \overset{\blacklozenge_2}{c_2} \to c_G \to \overset{\blacklozenge_1}{c_1} \to c_G$

$p_2 : c_I \to \overset{\lozenge_1}{c_1} \to c_I \to c_2 \to c_G \to c_3 \to c_I \to c_2 \to c_G$

$p_3 : c_I \to \overset{\lozenge_3}{c_3} \to c_I \to \overset{\lozenge_2}{c_2} \to c_I \to c_3 \to c_G$

Hence, there is no waiting time within P3s, and the optimal makespan is 32.

Proposition: For a given PPP C and β_{set}, the algorithm returns the optimal makespan.

Proof: The incompressible time to transport all passengers, according to a given β_{set} is $T = 4 \sum_{i \in \beta_{\text{set}}} d_i + 2 \sum_{i \in \{e', w'\}} d_i$. A theoretical optimal plan with this pattern repartition is a plan without any waiting point for any plane. The above algorithm gives the optimal distribution of the set of times into p. Then, if a plan can be constructed with such a makespan, it is optimal for the PPP and the repartition of patterns. As it exists a method to construct such a plan, we can conclude that the algorithm is optimal for the PPP C and β_{set}.   □

Complexity: Given a PPP, the worst case occurs when $w \subset e$ and $w_i \neq w_j$ if $i \neq j$. Hence, for each value of β there are $\binom{t-p}{\beta}$ possible β_{set}. As $0 \leq \beta \leq t-p$, we will perform 2^{t-p} *iterations* of the two step algorithm. A large upper-bound for the whole PPP set is hence $2^{t-p}\binom{n+t-1}{t}\binom{n+(t-p)-1}{t-p}$.

However, if an upper bound on β is given by $t-p$, a tighter upper-bound can be found as explained by the following example and due to the fact that the worst case situation for a PPP rarely occurs in the whole PPP set, the real number of iterations for a given instance is far from the above bound.

Example: Considering $C = (3,1,1)(2,1)$, the trivial upper bound is equal to two but actually, it is impossible to operate a P3 using the city C_2 since it is not in the tuple e.

3   ZenoSolver

ZenoSolver is a C++11 software dedicated to generate and exactly solve MultiZenoTravel instances. Firstly, it allows to tune every parameter in order to adjust the difficulty or to obtain different shapes of Pareto Fronts. In particular, vectors c and d are generated using two user-defined functions, f and g, such that $c_i = x_c f(i) + y_c$ and $d_i = x_d g(n-i) + y_d$, insuring that both objectives are conflicting. Second, ZenoSolver outputs the corresponding PDDL file[5], that can be directly used by any standard AI planner.

Finally, ZenoSolver computes the true Pareto Front using the algorithm described in Sect. 2, iterating over $E \times W$, storing for each value of the total cost the PPP with best makespan to date, without explicitly constructing the set of admissible PPPs. Using the Greedy domination, ZenoSolver implements a pruning method that checks if the current PPP is dominated by any other PPP already stored. As noted, the optimal makespan is lower or equal than the upper bound M_S, leading to an efficient pruning. Indeed, as PPPs are generated in an approximated increasing order [4], this avoids to iterate over the whole set to check the domination criterion.

Determining if the current PPP is dominated has complexity $O(h)$ where h is the number of different total costs. An obvious upper-bound for h is given by $(2t - p)(\max_i(c_i) - \min_i(c_i))$. However, in practice, S seems to have the same order of magnitude than the exact Pareto Front. In addition, S is the only structure kept in memory, thus, from this point of view, ZenoSolver turns out to be near-optimal regarding the memory usage (see Table 1).

3.1   Empirical Performances

Empirical complexity. The number of iterations is influenced by the number of PPPs but also by their structure. Indeed, increasing n does not significantly impact the average number of iterations per PPP since the upper-bound is 2^{t-p}.

---

[5] Planning Domain Definition Language, universally used now in AI Planning to describe domains and instances.

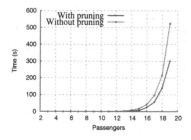 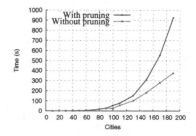

Fig. 2. Time function of t (left) or n (right) for $f(i) = g(i) = i$.

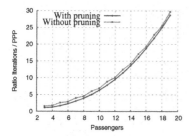 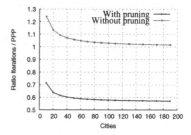

Fig. 3. Ratio of iterations over the number of PPPs, function of t (left) or n (right) for $f(i) = g(i) = i$.

On the opposite, increasing t leads to a dramatic growth of both the upper-bound and the average number of iterations per PPP. Figure 2, which displays the time vs t or n plots, confirms this remark: it requires the same CPU time for $t = 18$ than for $n = 165$.

Pruning or not pruning? The benefits of the pruning method strongly rely on the average number of iterations per PPP: Pruning becomes more efficient as t increases, as shown by Fig. 2. Furthermore, increasing n while pruning can degrade performances, even if there are less iterations than PPPs (Fig. 3 compared to Fig. 2). Note that the number of iterations follows the number of PPPs while increasing n, but explodes with t, which is in line with the previous remark.

Also, the efficiency pruning seems to be instance-dependent. Fixing n, t and p, different generating functions result in different numbers of iterations and CPU times, as demonstrated by Fig. 4. There are however some clear cases in favor of pruning, e.g. with $n = t = 9$: ZENOSOLVER requires 1.26×10^9 iterations and 2222 seconds without pruning. Using pruning, for $f(i) = \sqrt{i}$ and $g(i) = i$, it requires only 119000 iterations performed in 26 seconds, but 36000 iterations in 53 seconds with $f(i) = log(i)$ and $g(i) = \sqrt{i}$. In general, using concave generating functions leads to more optimistic conclusions regarding the benefits of pruning PPPs.

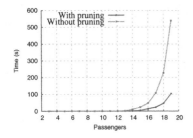 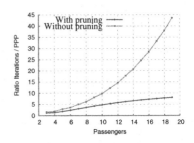

Fig. 4. Time and ratio for generating functions $f(i) = g(i) = log(i)$.

Table 1. Increasing simultaneously n and t with $f(i) = g(i) = i$.

n	t	p	PPP Size	Iterations	S Size	Front Size	Time (ms)
3	3	2	30	33	9	5	0
4	4	2	350	408	19	10	1
5	5	2	4410	6387	33	17	6
6	6	2	58212	109831	51	26	117
7	7	2	792792	1930385	73	37	2278
8	8	2	11042460	34648348	99	50	43572
9	9	2	156434850	630225670	129	65	1036772
10	10	2	2245709180	11600589455	163	82	20785211

3.2 Reference Large Instances

As mentioned in the introduction, the combinatorial multi-objective optimization domain suffers from a lack of benchmarks with a known Pareto Front but
also with a *concave* or *non-regular* shapes[6].

Even if anyone can generate different instances by tuning ZENOSOLVER parameters to obtain the desired front shape with accuracy, we identified some large
instances with totally different front shapes and complexities as displayed in
Fig. 5: They could be a basic set of representative instances for MULTIZENO-
TRAVEL, allowing fair comparisons between various solvers and approaches. Note
that more large instances with different complexities can be found on the website
of the Descarwin Project https://descarwin.lri.fr.

Table 2 gives the parameters used by ZENOSOLVER to build them, as well as
some statistics about their complexity. The choice of the generating functions is
purely empirical, guided by the fact that we would like to obtain mainly piecewise
concave fronts with uneven point distributions. This is why none of these fronts
is linear, though some seem to be at large scale (see detailed insets in some plots).
Also note the non-uniform distribution of the points on the Instances 3, and the

---

[6] In the context of discrete optimization, the word "concave" seems rather abusive.
However, we will call here concave parts of a Pareto front where all points are above
the segment made of the two extreme points, w.r.t. the direction of optimization.

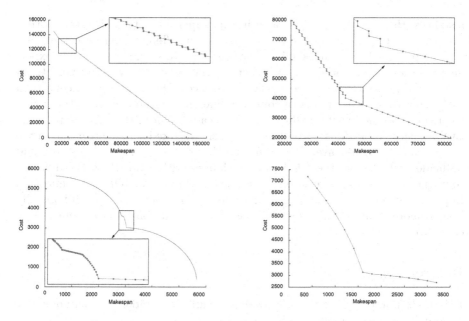

Fig. 5. True Pareto Fronts for the instances described by Table 2. Remember that these Pareto fronts are made of discrete points: the lines are visual helps to make the general shape clear.

Table 2. Large instances: parameters and generation statistics.

Inst.	n	t	p	Generating functions		Pareto#	h	PPP(k)	Iter.(k)	Time
1	20	6	2	$\frac{5}{2}i + \frac{(i \bmod 2)}{10}$	$\frac{5}{2}i + \frac{(i \bmod 2)}{10}$	409	4015	1568220	3317140	16h46
2	3	21	2			61	861	53	233	2006s
3	200	3	2	$\sqrt{i}$	$\sqrt{i}$	538	4963	270680	3906	1845s
4	8	26	25	$\sqrt{i}$	i	15	190	34176	60457	4240s

few Pareto points of the Instance 4 in spite of the complexity of this instance (26 persons), due to the small ratio $\frac{p}{t}$. The generating time strongly varies from some minutes up to hours and thus confirm dependency on the generating functions of the ZENOSOLVER complexity.

4   Conclusion and Perspectives

This paper has extended the MULTIZENOTRAVEL test suite in multi-objective AI planning. Furthermore, not only did we provide here a general approach to generate more complex Pareto fronts than in our previous work [2], but we also proposed here ZENOSOLVER, an exact solver that is provably able to exactly solve the multi-objective optimization problem (i.e., to identify the true Pareto front) for even very large instances. The complete code is publicly available at https:// descarwin.lri.fr, making it easy for everyone to generate his/her own benchmark instances. However, we also provided in this paper a few typical instances that exhibit very different shapes of Pareto Fronts, for different levels of complexity.

The proposed benchmark suite opens the floor to sound comparative experiments in a combinatorial domain where, as far as we know, no ground truth (i.e., true Pareto front) existed for large instances. On-going work is concerned with using these benchmarks to compare different multi-objective optimization algorithms. Preliminary results [7] have already confirmed that Pareto-based Evolutionary Algorithms outperform the basic weighted sum aggregation in the case of complex non-convex Pareto fronts. However, deeper experiments should be made with state-of-the-art decomposition algorithms in which the different components of the decomposition cooperate (e.g., from the MOEA/D family [6]). In particular, in the AI planning domain, using these benchmarks will hopefully lead to more sound comparisons between Pareto and non-Pareto planners (see e.g., [3,5]).

References

1. Igel, C., Hansen, N., Roth, S.: Covariance matrix adaptation for multi-objective optimization. Evol. Comput. **15**(1), 1–28 (2007)
2. Khouadjia, M.R., Schoenauer, M., Vidal, V., Dréo, J., Savéant, P.: Multi-objective AI planning: evaluating DAE$_{YAHSP}$ on a tunable benchmark. In: Purshouse, R.C., Fleming, P.J., Fonseca, C.M., Greco, S., Shaw, J. (eds.) EMO 2013. LNCS, vol. 7811, pp. 36–50. Springer, Heidelberg (2013)
3. Khouadjia, M.R., Schoenauer, M., Vidal, V., Dréo, J., Savéant, P.: Pareto-based multiobjective AI planning. In: Rossi, F. (ed.) IJCAI, pp. 2321–2328. AAAI Press, Menlo Park (2013)
4. Knuth, D.E.: The Art of Computer Programming, Generating All Tuples and Permutations. Addison-Wesley, Reading (2005)
5. Sroka, M., Long, D.: Exploring metric sensitivity of planners for generation of pareto frontiers. In: Kersting, K., Toussaint, M. (eds.) 6 STAIRS, pp. 306–317. IOS Press, Amsterdam (2012)
6. Zhang, Q., Li, H.: A multi-objective evolutionary algorithm based on decomposition. IEEE Trans. Evol. Comput. **11**(6), 712–731 (2007)
7. Quemy, A., Schoenauer, M., Vidal., V., Dréo, J., Savéant, P.: Solving large multizenotravel benchmarks with divide-and-evolve. In: Daenens, C., et al. (ed.) Proceedings of LION'9. Springer (2015, To appear)
8. Schoenauer, M., Savéant, P., Vidal, V.: Divide-and-evolve: a new memetic scheme for domain-independent temporal planning. In: Gottlieb, J., Raidl, G.R. (eds.) EvoCOP 2006. LNCS, vol. 3906, pp. 247–260. Springer, Heidelberg (2006)
9. Zitzler, E., Deb, K., Thiele, L.: Comparison of multiobjective evolutionary algorithms: empirical results. Evol. Comput. **8**(2), 173–195 (2000)

Upper and Lower Bounds on Unrestricted Black-Box Complexity of $\text{JUMP}_{n,\ell}$

Maxim Buzdalov[1(✉)], Mikhail Kever[1], and Benjamin Doerr[2]

[1] ITMO University, 49 Kronverkskiy av., Saint-Petersburg, Russia, 197101
{mbuzdalov,mikhail.kever}@gmail.com
[2] LIX, École Polytechnique, 91128 Palaiseau Cedex, France
doerr@lix.polytechnique.fr

Abstract. We analyse the unrestricted black-box complexity of $\text{JUMP}_{n,\ell}$ functions. For upper bounds, we present three algorithms for small, medium and extreme values of ℓ. We present a matrix lower bound theorem which is capable of giving better lower bounds than a general information theory approach if one is able to assign different types to queries and define relationships between them. Using this theorem, we prove lower bounds for JUMP separately for odd and even values of n. For several cases, notably for extreme JUMP, the first terms of lower and upper bounds coincide.

1 Introduction

To understand how evolutionary algorithms (and other black-box optimizers as well) behave when optimizing certain functions, it is possible to construct upper bounds (by constructing and studying various algorithms), as well as lower bounds (by studying how fast an algorithm can be in principle), which complement each other. Comparing these bounds allows to evaluate how good today's heuristics are and sometimes to construct better algorithms by learning from black-box [1].

Black-box complexity studies how many function evaluations are needed in expectation by an optimal black-box algorithm until it queries an optimum for the first time. As randomized search heuristics are black-box optimizers, black-box complexity of a problem gives a lower bound on the number of fitness evaluations of any search heuristic to solve this problem.

In this paper we consider optimization of functions mapping bit strings of fixed length to integers — the *pseudo-Boolean* functions. A famous class of such functions is ONEMAX — having a certain hidden bit string z of length n, for a bit string x of length n the function $\text{ONEMAX}_{n,z}(x)$ returns the number of bits coinciding both in x and z. JUMP, another popular class of functions, takes another parameter ℓ and zeroes out the values of ONEMAX for every string except z that are at the distance of at most ℓ from both z and its inverse.

© Springer International Publishing Switzerland 2015
G. Ochoa and F. Chicano (Eds.): EvoCOP 2015, LNCS 9026, pp. 209–221, 2015.
DOI: 10.1007/978-3-319-16468-7_18

A more formal definition of JUMP is as follows:

$$\text{JUMP}_{n,\ell,z} = \begin{cases} n & \text{if } \text{ONEMAX}_{n,z} = n \\ \text{ONEMAX}_{n,z} & \text{if } \ell < \text{ONEMAX}_{n,z} < n - \ell \\ 0 & \text{otherwise.} \end{cases}$$

Most of times, when z does not matter, we write just ONEMAX_n and $\text{JUMP}_{n,\ell}$. The special case of $\ell = \lfloor \frac{n}{2} \rfloor - 1$, which is the maximum possible ℓ that doesn't zero out the middle fitness values, is called *extreme* JUMP.

In this paper, we consider *unrestricted* black-box complexity (which was introduced in Droste et al. [4]) of the $\text{JUMP}_{n,\ell}$ problem. Another kind of black-box complexity, the *unbiased* black-box complexity, was considered for JUMP in Doerr et al. [2].

The rest of the paper is structured as follows. Section 2 is dedicated to the upper bounds on JUMP which are proven by giving the corresponding algorithms and discussing their complexity. In Sect. 3, the *matrix lower bound theorem*, which is somewhat similar to Theorem 2 from [4] but is able to produce better lower bounds, is described and proven. Section 4 describes lower bounds on JUMP which are constructed from the matrix theorem. Section 5 concludes.

2   Upper Bounds for Jump$_{n,\ell}$

Here, the upper bounds for JUMP are considered. In Sect. 2.1 several useful helper theorems are referenced or proven. Section 2.2 is dedicated to smaller ℓ, Sect. 2.3 is for larger ℓ, and Sect. 2.4 considers the case of extreme JUMP.

2.1   Helper Theorems

Theorem 1. *For sufficiently large n, for $t \geq \left(1 + \frac{4 \log_2 \log_2 n}{\log_2 n}\right) \frac{2n}{\log_2 n}$ and for an even $d \in [2; n]$ it holds that $\binom{n}{d} \left(\binom{d}{d/2} 2^{-d}\right)^t \leq 2^{-3t/4}$.*

Proof. This is proven in Doerr et al. [3] as Statement 8.   □

Theorem 2. *For sufficiently large n, for $\ell < n/2 - \sqrt{n} \log_2 n$ and for $x \in \{0,1\}^n$ taken uniformly at random the probability for $\text{JUMP}_{n,\ell}(x)$ to be zero is at most $2e^{-2(\log_2 n)^2}$.*

Proof. The value of $\text{ONEMAX}(x)$ for a random x has a binomial distribution with parameters n and $p = 1/2$. From Hoeffding's inequality [5], for $k \leq np$, the distribution function for binomial distribution $F_{n,p}(k)$ is bound from above by $e^{-2\frac{(np-k)^2}{n}}$. As a consequence, the probability for $\text{JUMP}_{n,\ell}(x)$ to be zero is at most $2F_{n,1/2}(l) \leq 2e^{-2\frac{(n/2-l)^2}{n}} \leq 2e^{-2\frac{(\sqrt{n}\log_2 n)^2}{n}} = 2e^{-2(\log_2 n)^2}$.   □

Theorem 3. *Assume that n is sufficiently large and $\ell < n/2 - \sqrt{n}\log_2 n$. Let $z \in \{0,1\}^n$, and X be a set of $t \geq \left(1 + \frac{4\log_2\log_2 n}{\log_2 n}\right)\frac{2n}{\log_2 n}$ elements from $\{0,1\}^n$ chosen randomly using uniform distribution and mutually independently. The probability that there exists an $y \in \{0,1\}^n$ such that $y \neq z$ and $\textsc{Jump}_{n,\ell,z}(x) = \textsc{Jump}_{n,\ell,y}(x)$ for all $x \in X$, is at most $2^{-t/4}$.*

Proof. Let's define A_d as a set of points which differ from z in exactly d positions where $0 \leq d \leq n$.

We say that a point $y \in \{0,1\}^n$ *agrees* with $x \in X$ if $\textsc{Jump}_{n,\ell,z}(x) = \textsc{Jump}_{n,\ell,y}(x)$. This means that either $\textsc{Jump}_{n,\ell,z}(x) = 0$ or $\textsc{OneMax}_{n,y}(x) = \textsc{OneMax}_{n,z}(x)$. The probability of the former does not exceed $2e^{-2(\log_2 n)^2}$ by Theorem 2. The latter holds iff x and y (as well as x and z) differ in exactly half of the d bits in which y and z differ. To sum up, if $y \in A_d$, the probability for y to agree with a random x is at most $2e^{-2(\log_2 n)^2}$ for an even d and at most $2e^{-2(\log_2 n)^2} + \binom{d}{d/2}2^{-d}$ for an odd d. As for large enough n it holds that $2e^{-2(\log_2 n)^2} \leq \left(2^{1/4} - 1\right)\binom{d}{d/2}2^{-d}$, the latter is at most $2^{1/4}\binom{d}{d/2}2^{-d}$.

Let p be the probability that there exists an $y \in \{0,1\}^n \setminus \{z\}$ such that y agrees with all $x \in X$. Then the following holds:

$$p = Pr\left(\bigcup_{y\in\{0,1\}^n\setminus\{z\}}\bigcap_{x\in X} y \text{ agrees with } x\right)$$

$$\leq \sum_{y\in\{0,1\}^n\setminus\{z\}} Pr\left(\bigcap_{x\in X} y \text{ agrees with } x\right)$$

$$= \sum_{d=1}^{n}\sum_{y\in A_d}\prod_{x\in X} Pr(y \text{ agrees with } x)$$

$$\leq \sum_{d \text{ even}}\binom{n}{d}\left(2^{1/4}\binom{d}{d/2}2^{-d}\right)^t + \sum_{d \text{ odd}}\binom{n}{d}\left(2e^{-2(\log_2 n)^2}\right)^t$$

$$= \sum_{d \text{ even}}\binom{n}{d}\left(2^{1/4}\binom{d}{d/2}2^{-d}\right)^t + 2^{n-1}\left(2e^{-2(\log_2 n)^2}\right)^t.$$

After applying Theorem 1, we get that:

$$p \leq \frac{n+1}{2}2^{t/4}2^{-3t/4} + 2^{n-1+t}e^{-2t(\log_2 n)^2},$$

which is less than $2^{-t/4}$ for sufficiently large n. $\square$

2.2 Upper Bound for Smaller ℓ

Theorem 4. *If $\ell < n/2 - \sqrt{n}\log_2 n$, the unrestricted black-box complexity of $\textsc{Jump}_{n,\ell}$ is at most $(1 + o(1))\frac{2n}{\log_2 n}$, where $o(1)$ is measured relative to n.*

Proof. We use the same algorithm which is used in [3] for proving the lower bound for ONEMAX. We select randomly and independently t queries such that $t \geq \left(1 + \frac{4\log_2\log_2 n}{\log_2 n}\right)\frac{2n}{\log_2 n}$ and check if there exists a single optimum z which agrees with all these queries (a query q with an answer a agrees with an optimum z if $\text{JUMP}_{n,\ell,z}(q) = a$). The complexity of one iteration equals to t and the probability of not finding an optimum is at most $2^{-t/4}$ by Theorem 3. Thus the complexity of the algorithm is at most $\frac{t}{1-2^{-t/4}} = (1 + o(1))\frac{2n}{\log_2 n}$. □

2.3 Upper Bound for Larger ℓ

For bigger l, the $\text{JUMP}_{n,\ell}$ problem can be solved by reduction to smaller JUMP problems for which the algorithm for the previous section suffices.

Theorem 5. *For $\frac{n}{2} - \sqrt{n}\log_2 n \leq \ell < \left\lfloor\frac{n}{2}\right\rfloor - 1$ the unrestricted black-box complexity of $\text{JUMP}_{n,\ell}$ is at most $(1 + o(1))\frac{n}{\log_2(n-2\ell)}$ where $o(1)$ is measured when $(n - 2\ell) \to \infty$.*

Proof. Assume that $k = \left\lfloor\frac{n}{2}\right\rfloor - \ell - 1 \neq 0$. We reduce our problem to $\text{JUMP}_{s,\frac{s}{2}-k-1}$ where $\sqrt{s}\log_2 s < k$. The algorithm is outlined at Fig. 1.

First the algorithm finds a maximum even s such that $\sqrt{s}\log_2 s < k$, which would allow applying Theorem 4 for solving $\text{JUMP}_{s,\frac{s}{2}-k-1}$. After that, the algorithm finds a string $x \in \{0,1\}^n$ with exactly $\left\lfloor\frac{n}{2}\right\rfloor$ correct bits using random queries. The probability that $\text{JUMP}_{n,\ell}$ is equal to $\left\lfloor\frac{n}{2}\right\rfloor$ for a random query is $2^{-n}\binom{n}{\lfloor n/2\rfloor}$ which is $\Theta\left(\frac{1}{\sqrt{n}}\right)$ by Stirling's formula. This means that the string x can be found in $\Theta(\sqrt{n})$ queries.

After finding the x, the algorithm splits all bit indices into sets of size s (except for probably one) in such a way that in each set exactly half of bits coincide with those in the answer. This is done in lines 8–15 at Fig. 1, where b_i is the i-th such set. B, the set of yet undistributed bits, always contains indices of which exactly $|B|/2$ indices correspond to correctly guessed bits.

To do that, the algorithm generates random subsets of size s and checks each of them if it contains exactly $\frac{s}{2}$ correct bits, which is done by flipping the bits from the chosen subset and checking whether the fitness function returns $\left\lfloor\frac{n}{2}\right\rfloor$. If $|B| = m$, the probability of correct selection is:

$$p = \binom{\lfloor m/2\rfloor}{s/2}\binom{\lceil m/2\rceil}{s/2}\binom{m}{s}^{-1} = \frac{\lfloor m/2\rfloor!\,\lceil m/2\rceil!\,s!\,(m-s)!}{\lfloor(m-s)/2\rfloor!\,\lceil(m-s)/2\rceil!\,(s/2)!}$$

$$= \binom{s}{s/2}\binom{m-s}{\lfloor(m-s)/2\rfloor}\binom{m}{\lfloor m/2\rfloor}^{-1} = \Theta\left(\frac{\frac{2^s}{\sqrt{s}}\frac{2^{m-s}}{\sqrt{m-s}}}{\frac{2^m}{\sqrt{m}}}\right)$$

$$= \Theta\left(\frac{\sqrt{m}}{\sqrt{s}\sqrt{m-s}}\right) = \Omega\left(\frac{1}{\sqrt{s}}\right).$$

This gives an $O(\sqrt{s})$ bound for one subset selection and an $O(n/\sqrt{s})$ bound on entire process of finding subsets.

```
 1: function LARGEJUMP(n, ℓ, f ∈ JUMPₙ,ℓ)
 2:     k ← ⌊n/2⌋ - ℓ - 1
 3:     s ← max{w|√w log₂ w < k; w is even}
 4:     τ ← ⌈n/s⌉
 5:     repeat
 6:         x ← UNIFORM()
 7:     until f(x) = ⌊n/2⌋
 8:     B ← [1;n]
 9:     for i ∈ [1;τ) do
10:         repeat
11:             bᵢ ← CHOOSESUBSETRANDOMLYUNIFORMLY(B, s)
12:         until f(FLIPBITS(x, bᵢ)) = ⌊n/2⌋
13:             B ← B \ bᵢ
14:     end for
15:     bτ ← B
16:     ω₀ ← x
17:     for i ∈ [1;τ] do
18:         αᵢ ← SMALLJUMPPROJECTION(bᵢ, x)
19:         ωᵢ ← SETATPOSITIONS(bᵢ, ωᵢ₋₁, αᵢ)
20:     end for
21:     return ωτ
22: end function
```

Fig. 1. Algorithm for $\text{JUMP}_{n,\ell}$ with $\frac{n}{2} - \sqrt{n} \log_2 n \le \ell < \lfloor \frac{n}{2} \rfloor - 1$

Next, the algorithm optimizes separately bits from each of the subsets b_i using the algorithm for small JUMP from Theorem 4 (lines 17–20 at Fig. 1). If every query for a subproblem on bits from b_i is forwarded to the main function f with all bits not from b_i taken from x, the resulting subproblem becomes exactly a $\text{JUMP}_{|b_i|, \frac{|b_i|}{2} - k - 1}$ problem with the following corrections:

- from all nonzero answers, a value of $\lfloor \frac{n - |b_i|}{2} \rfloor$ needs to be subtracted;
- at the optimum of the subproblem, zero will be returned.

The latter correction, however, does not change the algorithm very much, because the algorithm from Theorem 4 doesn't actually query the optimum point. The line 19 from Fig. 1 collects the partial answers one by one: it sets the bits of a_i at the corresponding positions from b_i to the previous partial answer ω_{i-1} and returns the updated value.

The complexity of the algorithm can be expressed as (here $n = qs + r$, $(0 < r \le s)$):

$$O(\sqrt{n}) + O\left(\frac{n}{\sqrt{s}}\right) + q(2 + o_s(1))\frac{s}{\log_2 s} + (2 + o_r(1))\frac{r}{\log_2 r} = \frac{(2 + o_s(1))n}{\log_2 s}.$$

However, due to choice of s, it holds that $\log_2 s = (2 + o_k(1)) \log_2 k$, which finally results in $\frac{(1 + o_{n-2\ell}(1))n}{\log_2(n - 2\ell)}$. $\qquad\qquad \square$

2.4 Upper Bound for Extreme Jump

The algorithm from Theorem 5 cannot be applied to the case of extreme JUMP, because $\lfloor \frac{n}{2} \rfloor - 1 - k = 0$. In this case we have to use another algorithm, which will be given in the proof of the following theorem.

Theorem 6. *The unrestricted black-box complexity of the extreme* JUMP *is at most* $n + \Theta(\sqrt{n})$.

Proof. As described in previous theorems, one can find a point x, such that $f(x) = \lfloor \frac{n}{2} \rfloor$, in $\Theta(\sqrt{n})$ queries. After that, if one flips two bits, the value of f remains the same iff one of these bits was correct and the other was not.

Once x is found, the algorithm tests $f(x \oplus 10^{i-1}10^{n-i-1})$ for all $i \in [2; n]$, and if it equals $\lfloor \frac{n}{2} \rfloor$, the value of b_i is set to zero, otherwise to one. This results in $n - 1$ queries. After that, if the first bit is correct, then $0b_2 \dots b_n$ is the answer, otherwise its inverse is the answer. One has to make a single query to $f(0b_2 \dots b_n)$ to find which one is true. The complexity of this algorithm is $n + \Theta(\sqrt{n})$. $\square$

3   The Matrix Lower Bound Theorem

In this section we present a new theorem which is similar to Theorem 2 from [4] except that the nodes corresponding to queries are required to be split in several types.

Theorem 7. *Let S be the search space of an optimization problem, and for each $s \in S$ there exists an instance such that s is a unique optimum. Let each query has one of T types, such that for any query q of the i-th type the following holds:*

- *there is exactly one answer to the query q which means that q is an optimum;*
- *there are at most $A_{i,j}$ answers such that the next query after such answer belongs to the j-th type.*

Define $B_{i,j}, 1 \le i, j \le T + 2$, to be a matrix such that:

- *$B_{i,j} = A_{j,i}$ for $1 \le i, j \le T$ (note the transposition);*
- *$B_{T+1,j} = 1$ for $1 \le j \le T + 1$;*
- *$B_{T+2,j} = 1$ for $1 \le j \le T + 2$;*
- *$B_{i,j} = 0$ otherwise.*

Let the first ever query in the optimization process be of type 1. Define $V(d) = B^d \cdot (1, 0, \dots, 0)^T$ be a vector, $C(d) = V(d)_{T+1}$, $S(d) = V(d)_{T+2}$. Then the following statements are true:

1. *$C(d)$ is the maximum total number of possible queries with depth in $[1; d]$, where depth of a root is equal to one.*
2. *The lower bound on average depth of N nodes is $d + 1 - \frac{S(d)}{N}$ where d is an integer such that $C(d) \le N \le C(d + 1)$.*
3. *The unrestricted black-box complexity of the considered optimization problem is not less than the lower bound on average depth of $|S|$ nodes.*

Proof. According to Yao's minimax principle [6], the expected runtime of a randomized algorithm on any input is not less than the average runtime of the best deterministic algorithm over all possible inputs. Thus we construct a lower bound on complexity of a randomized algorithm by constructing a lower bound on the average performance of any deterministic algorithm over all possible inputs. A deterministic algorithm can be represented as a (rooted) decision tree with nodes corresponding to queries and arcs going downwards corresponding to answers to these queries. A total lower bound on the average performance of deterministic algorithms, just as in [4], is done by assigning $|S|$ different queries to different nodes of a tree such that their average depth is minimized, and then by considering all such trees and taking a minimum over them.

It should be noted that, if a (fixed) set of queries is to be assigned to nodes of a (fixed) rooted tree such that the average depth of these queries is minimized, an optimal assignment can be constructed in a greedy way: each query should be assigned to a free node with the minimum possible depth. Assume that an optimal assignment does not use at least one node a with depth d while using at least one node b with depth $d' > d$. Then one can move a query from the node b to the node a, which makes the average depth decrease, so the initial assignment is, in fact, not optimal.

Next we show that, in order to minimize the average depth, one needs to consider only the complete tree, that is, a tree where for any query of the i-th type, for any j there are exactly $A_{i,j}$ answers, each leading to a query of the j-th type. Indeed, if an optimal assignment can be done for an incomplete tree, it can be done for the complete tree as well, because all the nodes of any incomplete tree are preserved in the complete tree.

For a complete tree with the constraints determined by the matrix A (as specified in the theorem's statement) and with the root vertex of type 1, the number of vertices of type i and depth d (the root has the depth equal to 1) is exactly $\left((A^T)^{d-1} \cdot (1, 0, \ldots, 0)^T \right)_i$. In the matrix B, the next-to-last row is designed to collect the sum of all numbers of vertices at all previous depths (which is exactly how $C(d)$ is defined), and the last row, in a similar manner, collects $S(d)$ — the sum of $C(i)$'s for all $1 \leq i \leq d$. In a more explicit way, $S(d)$ can be expressed as:

$$S(d) = \sum_{i=1}^{d} C(i) \cdot (d + 1 - i),$$

so the expression $C(d) \cdot (d+1) - S(d)$ is actually the sum of depths of all vertices up to the depth d:

$$C(d)(d + 1) - S(d) = \sum_{i=1}^{d} C(i) \cdot i,$$

and the expression $d + 1 - \frac{S(d)}{C(d)}$ is thus exactly the average depth of all such vertices.

If we consider arbitrary integer N, we can find an integer d such that $C(d) \leq N \leq C(d + 1)$. In this case, the total sum of depths of the first $C(d)$ vertices is

$C(d) \cdot (d+1) - S(d)$, and the next $N - C(d)$ vertices have the depth of $d+1$. The average depth is thus:

$$d_{avg}(N) = \frac{C(d) \cdot (d+1) - S(d) + (d+1) \cdot (N - C(d))}{N} = d + 1 - \frac{S(d)}{N}. \qquad \square$$

It is difficult to use this theorem straightaway, because the lower bound on the average depth of N vertices is not defined only in terms of N and the matrix A, but additionally requires to find which depth d fulfils $C(d) \leq N \leq C(d+1)$. However, for several common usages it is possible to make it more convenient.

Theorem 8. *If there is only one type of queries in Theorem 7, and $A_{1,1} = k$ such that $k \geq 2$, then for the search space S the lower bound on the average depth is at least $\lfloor \log_k(1 + |S|(k-1)) \rfloor - \frac{1}{k-1}$.*

Proof. The value of $B^d \cdot (1,0,0)^T$ yields the following result (intermediate computations omitted):

$$\begin{pmatrix} k & 0 & 0 \\ 1 & 1 & 0 \\ 1 & 1 & 1 \end{pmatrix}^d \cdot \begin{pmatrix} 1 \\ 0 \\ 0 \end{pmatrix} = \begin{pmatrix} k^d \\ \frac{k^d-1}{k-1} \\ \frac{k^{d+1}-k-d(k-1)}{(k-1)^2} \end{pmatrix}.$$

One can see that $C(d) = \frac{k^d-1}{k-1}$ and $S(d) = \frac{k^{d+1}-k-d(k-1)}{(k-1)^2}$.

Consider an equality $N = C(d) = \frac{k^d-1}{k-1}$. It follows that:

$$d(N) = \log_k(1 + N(k-1)).$$

As for a given N we need to find an integer d such that $C(d) \leq N < C(d+1)$, we need to round it down: $d = \lfloor d(N) \rfloor$.

Note that, if $d \geq 1$ and $k \geq 1$, $S(d)$ grows when d grows, as $S(d)' > 0$.

The expression for a lower bound on the average depth of N queries is at most:

$$d_{avg}(N) = \lfloor d(N) \rfloor + 1 - \frac{S(\lfloor d(N) \rfloor)}{N} \geq \lfloor d(N) \rfloor + 1 - \frac{S(d(N))}{N}$$

$$\geq \lfloor \log_k(1 + N(k-1)) \rfloor - \frac{1}{k-1}. \qquad \square$$

Note that the classical result from [4], the $\lfloor \log_{k+1} N \rfloor - 1$ lower bound, is actually not greater than the given bound. Indeed, for $k \geq 2$:

$$\log_k(1 + N(k-1)) - \log_{k+1} N > \log_k(N(k-1)) - \log_{k+1} N$$

$$= \log_k N - \log_{k+1} N + \log_k(k-1) > \log_k N - \log_{k+1} N > 0.$$

For the case of $k = 1$, the lower bound is even stronger.

Theorem 9. *If there is only one type of queries in Theorem 7, and $A_{1,1} = 1$, then for the search space S the lower bound on the average depth is at least $(|S| + 1)/2$.*

Proof. In this case one can show that $C(d) = d$ and $S(d) = \frac{d^2+d}{2}$. The average depth for N is $N + 1 - \frac{N^2+N}{2N} = N + 1 - (N+1)/2 = (N+1)/2$. $\qquad \square$

4   Lower Bounds for Jump$_{n,\ell}$

First, let's apply Theorem 8 immediately to the JUMP problem.

Theorem 10. *For any n and $\ell < n/2$, the unrestricted black-box complexity of* JUMP$_{n,\ell}$ *is at least* $\lfloor \log_{n-2\ell}(1 + 2^n(n - 2\ell - 1)) \rfloor - \frac{1}{n-2\ell-1}$.

Proof. In JUMP$_{n,\ell}$, the search space has a size of 2^n. There are $n - 2\ell + 1$ possible answers to a query, but one of them terminates the search process immediately, so $k = n - 2\ell$. The result follows straightaway from Theorem 8.

Theorem 11. *The unrestricted black-box complexity of extreme* JUMP *for even n is at least $n - 1$.*

Proof. Follows from Theorem 10 by assuming $n - 2\ell = 2$.

The presented bounds are already an improvement over the currently known bounds (say, $\frac{n}{\log_2 3}$ for extreme JUMP and even n, as follows from [4]). However, for odd n Theorem 10 reports $\lfloor \log_3(1 + 2^{n+1}) \rfloor - 1/2$, which is still quite far away from the best known algorithms. Fortunately, the JUMP problem possesses a particular property, which can be used to refine the lower bounds using Theorem 7 with *two types* of queries.

Theorem 12. *For* JUMP$_{n,\ell}$, *define an answer to the query to be* non-trivial *if it is neither 0 nor n. After receiving the first non-trivial answer for every subsequent query it is possible to determine* a priori *the parity of any non-trivial answer.*

Proof. Consider the optimum and a query. We introduce the following values:

- q_{00}: number of positions with zeros in both the optimum and the query;
- q_{01}: number of positions with zeros in the optimum and ones in the query;
- q_{10}: number of positions with ones in the optimum and zeros in the query;
- q_{11}: number of positions with ones in both the optimum and the query.

The number of zeros in the optimum modulo 2, which is $q_{00} \oplus q_{01}$, is fixed. The number of ones in the query modulo 2 is $q_{01} \oplus q_{11}$, and the answer to the query modulo 2 is $q_{00} \oplus q_{11}$. The following equality holds:

$$(q_{01} \oplus q_{11}) \oplus (q_{00} \oplus q_{11}) = q_{00} \oplus q_{01},$$

which means that the parity of the non-trivial answer is uniquely determined by the parity of the number of ones in the query.

As a result, if an algorithm receives the first non-trivial answer, all subsequent queries will provably have fewer possible answers.   □

Using Theorem 12, we can define two types of queries to use with Theorem 7, namely, the queries happened before and after a non-trivial answer.

Theorem 13. *The unrestricted black-box complexity of* $\text{JUMP}_{n,\ell}$ *for odd* n *is at least:*

$$\left\lfloor \log_{\frac{n-2\ell+1}{2}} \left(2^{n-2}(n-2\ell-1)+1\right)\right\rfloor - \frac{2}{n-2\ell-1}.$$

Proof. For odd n there are $n - 2\ell + 1 = 2k + 2$ possible answers: one answer equal to 0, one answer equal to n and k pairs of non-trivial answers. For the Theorem 7, the first type of queries has $2k + 1$ non-terminating answers, and the second type of queries, which occurs after one of $2k$ non-trivial answers is received from a query of the first type, has only $k + 1$ non-terminating answers. The value of $B^d \cdot (1,0,0,0)^T$ is thus:

$$B^d \cdot \begin{pmatrix} 1 \\ 0 \\ 0 \\ 0 \end{pmatrix} = \begin{pmatrix} 1 & 0 & 0 & 0 \\ 2k & k+1 & 0 & 0 \\ 1 & 1 & 1 & 0 \\ 1 & 1 & 1 & 1 \end{pmatrix}^d \cdot \begin{pmatrix} 1 \\ 0 \\ 0 \\ 0 \end{pmatrix} = \begin{pmatrix} 1 \\ 2\left((k+1)^d - 1\right) \\ \frac{2(k+1)^d - dk - 2}{k} \\ 2(k+1)^{d+1} - \frac{(dk+2)^2 + dk^2 + 4k}{k^2} \end{pmatrix}.$$

A problem of defining d in terms of N is more difficult this time: as $C(d) = \frac{2(k+1)^d - dk - 2}{k}$, the equality $N = C(d)$ cannot be easily solved in terms of d. Instead, we introduce a function $d(N)$ such that the following equality holds:

$$N = \frac{2(k+1)^{d(N)} - d(N)k - 2}{k}.$$

We find the lower bound on the average depth $d_{avg}(N)$, keeping in mind that $S(d)$ grows as d grows and that $d(N) \geq 1$ for $N \geq 1$:

$$d_{avg}(N) = \lfloor d(N)\rfloor + 1 - \frac{S(\lfloor d(N)\rfloor)}{N} \geq \lfloor d(N)\rfloor + 1 - \frac{S(d(N))}{N}$$

$$= \lfloor d(N)\rfloor + 1 - \frac{2(k+1)^{1+d(N)} - \frac{(d(N)k+2)^2 + d(N)k^2 + 4k}{k^2}}{\frac{2(k+1)^{d(N)} - d(N)k - 2}{k}}$$

$$= \lfloor d(N)\rfloor + 1 - \frac{\frac{2(k+1)^{1+d(N)} - d(N)k^2 - d(N)k - 2k - 2 - \frac{d(N)k^2(d(N)-1)}{2}}{k}}{\frac{2(k+1)^{1+d(N)} - d(N)k^2 - d(N)k - 2k - 2}{k+1}}$$

$$\geq \lfloor d(N)\rfloor + 1 - \frac{k+1}{k} = \lfloor d(N)\rfloor - \frac{1}{k}.$$

We can also obtain a good lower bound on $d(N)$ by throwing out the $d(N)k$ part in the definition of $d(N)$ above, which leads to $d(N) > \log_{k+1}\left(\frac{Nk}{2} + 1\right)$. Together, $d_{avg}(N) \geq \left\lfloor \log_{k+1}\left(\frac{Nk}{2} + 1\right)\right\rfloor - \frac{1}{k}$. For $\text{JUMP}_{n,\ell}$, it holds that $N = 2^n$ and $2k + 2 = n - 2\ell + 1$, which constitutes:

$$\left\lfloor \log_{\frac{n-2\ell+1}{2}} \left(2^{n-2}(n-2\ell-1)+1\right)\right\rfloor - \frac{2}{n-2\ell-1}. \qquad \square$$

Theorem 14. *The unrestricted black-box complexity of extreme* JUMP *for odd* n *is at least* $n - 2$.

Proof. For extreme JUMP and odd n, $n - 2\ell + 1 = 4$. Then from Theorem 13 it follows that the lower bound is at least:

$$\lfloor \log_2 \left(2^{n-2} \cdot 2 + 1\right) \rfloor - \frac{2}{2} = \lfloor \log_2 \left(2^{n-1} + 1\right) \rfloor - 1 \geq n - 2.$$

Theorem 15. *The unrestricted black-box complexity of* JUMP$_{n,\ell}$ *for even* n *is at least:*

$$\left\lfloor \log_{\frac{n-2\ell+2}{2}} \left(1 + 2^{n-1} \frac{(n - 2\ell)^2}{n - 2\ell - 1}\right) \right\rfloor - \frac{2}{n - 2\ell}.$$

Proof. For even n there are $n - 2\ell + 1 = 2k + 3$ possible answers $(k \geq 0)$: one answer equal to 0, one answer equal to n, one answer equal to $n/2$ and k more pairs of non-trivial answers. For Theorem 7, the first type of queries has $2k + 2$ non-terminating answers, and the second type of queries can have either $k + 1$ or k non-terminating answers, depending on the parity of the number of ones in a query. As we cannot predict the parity for all possible algorithms, the maximum number of queries is limited to $k + 1$. The matrix B has the following form:

$$B = \begin{pmatrix} 1 & 0 & 0\,0 \\ 2k + 1 & k + 2 & 0\,0 \\ 1 & 1 & 1\,0 \\ 1 & 1 & 1\,1 \end{pmatrix}.$$

We omit the intermediate computations and just state that:

$$C(d) = \frac{(2k + 1)(k + 2)^d - dk^2 - dk - 2k - 1}{(k + 1)^2}$$

$$S(d) = \frac{(k + 2)^d (2k^2 + 5k + 2) - \frac{d^2k^3 + dk^3 + 2d^2k^2 + 6dk^2 + 4k^2 + 2d + d^2k + 7dk + 10k + 4}{2}}{(k + 1)^3}.$$

Following the same approach as in the proof of Theorem 13, we define $d(N)$ such that $C(d(N)) = N$ and produce the following lower bound:

$$d_{avg}(N) \geq \lfloor d(N) \rfloor - \frac{1}{k + 1}.$$

The lower bound on $d(N)$ can be achieved from the value of $C(d)$ by throwing out the $dk^2 + dk$ part, which yields:

$$d(N) \geq \log_{k+2} \left(1 + \frac{(k + 1)^2 N}{2k + 1}\right)$$

and, together:

$$d_{avg}(N) \geq \left\lfloor \log_{k+2} \left(1 + \frac{(k + 1)^2 N}{2k + 1}\right) \right\rfloor - \frac{1}{k + 1}.$$

Substitution of N with 2^n and $2k + 2$ with $n - 2\ell$ proves the theorem.   □

Note that Theorem 15 does not improve the bound for extreme JUMP and even n — it remains equal to $n - 1$ when one sets $k = 0$ — because in this case the number of possible answers does not change after receiving the first non-trivial answer.

5   Conclusion

New black-box algorithms for solving $\text{JUMP}_{n,\ell}$ problem are presented, giving the following upper bounds:

- for $\ell < n/2 - \sqrt{n}\log_2 n$: $\frac{2n(1+o(1))}{\log_2 n}$, where $o(1)$ is measured when $n \to \infty$;
- for $n/2 - \sqrt{n}\log_2 n \le \ell < \lfloor \frac{n}{2} \rfloor - 1$: $\frac{n(1+o(1))}{\log_2(n-2\ell)}$, where $o(1)$ is measured when $n - 2\ell \to \infty$;
- for $\ell = \lfloor \frac{n}{2} \rfloor - 1$: $n + \Theta(\sqrt{n})$.

A new theorem for constructing lower bounds on unrestricted black-box complexity of problems is proposed. The underlying idea is that influence of particular answers to queries to all subsequent queries can be formalized by assigning a type to each query and writing the relations in a form of a matrix. Several following steps for constructing the lower bounds are automated and can be performed using tools like Wolfram Alpha. We hope that this theorem can be used to obtain better lower bounds in other problems.

Using the proposed theorem, the lower bounds for $\text{JUMP}_{n,\ell}$ are updated:

- for even n: $\left\lfloor \log_{\frac{n-2\ell+2}{2}} \left(1 + 2^{n-1} \frac{(n-2\ell)^2}{n-2\ell-1} \right) \right\rfloor - \frac{2}{n-2\ell} \ge \frac{n}{\log_2 \frac{n-2\ell+2}{2}} - 1$;
- for odd n: $\left\lfloor \log_{\frac{n-2\ell+1}{2}} \left(1 + 2^{n-2}(n - 2\ell - 1) \right) \right\rfloor - \frac{2}{n-2\ell-1} \ge \frac{n-1}{\log_2 \frac{n-2\ell+1}{2}} - 1$.

In particular, for extreme JUMP the lower bounds become equal to $n - 1$ for even n and $n - 2$ for odd n. This means that the quotients at the first term of lower and upper bounds coincide. In the case of large, but not extreme ℓ, these quotients seem to coincide as well, however, the $(1 + o(1))$ multiple can hide as much as $\left(\log_2(n - 2\ell)/\log_2 \frac{n-2\ell+1}{2} \right)$, which makes it hard to see exactly.

This work was partially financially supported by the Government of Russian Federation, Grant 074-U01.

References

1. Doerr, B., Doerr, C., Ebel, F.: Lessons from the black-box: fast crossover-based genetic algorithms. In: Proceedings of Genetic and Evolutionary Computation Conference, pp. 781–788 (2014)
2. Doerr, B., Doerr, C., Kötzing, T.: Unbiased black-box complexities of jump functions. http://arxiv.org/abs/1403.7806v2
3. Doerr, B., Johannsen, D., Kötzing, T., Lehre, P.K., Wagner, M., Winzen, C.: Faster black-box algorithms through higher arity operators. In: Proceedings of Foundations of Genetic Algorithms, pp. 163–172 (2011)

4. Droste, S., Jansen, T., Wegener, I.: Upper and lower bounds for randomized search heuristics in black-box optimization. Theory Comput. Syst. **39**(4), 525–544 (2006)
5. Hoeffding, W.: Probability inequalities for sums of bounded random variables. J. Am. Stat. Assoc. **58**(301), 13–30 (1963)
6. Yao, A.C.C.: Probabilistic computations: toward a unified measure of complexity. In: 18th Annual Symposium on Foundations of Computer Science, pp. 222–227 (1977)

Using Local Search to Evaluate Dispatching Rules in Dynamic Job Shop Scheduling

Rachel Hunt[1]([✉]), Mark Johnston[1], and Mengjie Zhang[2]

[1] School of Mathematics, Statistics and Operations Research, Victoria University
of Wellington, PO Box 600, Wellington, New Zealand
[2] School of Engineering and Computer Science, Victoria University of Wellington,
PO Box 600, Wellington, New Zealand
huntrach1@myvuw.ac.nz, {mark.johnston,mengjie.zhang}@vuw.ac.nz

Abstract. Improving scheduling methods in manufacturing environments such as job shops offers the potential to increase throughput, decrease costs, and therefore increase profit. This makes scheduling an important aspect in the manufacturing industry. Job shop scheduling has been widely studied in the academic literature because of its real-world applicability and difficult nature. Dispatching rules are the most common means of scheduling in dynamic environments. We use genetic programming to search the space of potential dispatching rules. Dispatching rules are often short-sighted as they make one instantaneous decision at each decision point. We incorporate local search into the *evaluation* of dispatching rules to assess the quality of decisions made by dispatching rules and encourage the dispatching rules to make good local decisions for effective overall performance. Results show that the inclusion of local search in evaluation led to the evolution of dispatching rules which make better decisions over the local time horizon, and attain lower total weighted tardiness. The advantages of using local search as a tie-breaking mechanism are not so pronounced.

1 Introduction

Scheduling is a difficult yet important decision making process in manufacturing and service industries [19], involving allocating resources to complete a set of tasks to optimise some objective subject to constraints [7]. Job shop scheduling (JSS) problems have been extensively studied in the academic literature over the past 60 years [7,20], due mainly to the real-world applicability (there are many thousands of job shops world-wide [13]) and the computational complexity of scheduling in such environments (JSS is known to be *NP*-hard [3]).

In JSS problems, a set of jobs must be processed through a set of machines. Each job consists of a sequence of operations which must be processed in order, where each operation has a specified machine and processing time. The aim is to schedule jobs at each machine to optimise a measure of delivery speed or customer satisfaction. In *dynamic* JSS, jobs arrive according to a stochastic process; no information is available about a job until it is present in the shop system.

© Springer International Publishing Switzerland 2015
G. Ochoa and F. Chicano (Eds.): EvoCOP 2015, LNCS 9026, pp. 222–233, 2015.
DOI: 10.1007/978-3-319-16468-7_19

This makes the scheduling task very difficult, and optimisation methods which construct complete schedules in advance are not able to be used. Dispatching rules (DRs) are mathematical functions of attributes of the jobs, machines and the entire job shop. They have been applied consistently to complex scheduling problems as they provide good solutions in real-time [13]. DRs are popular because of their low time complexity and easy interpretability as well as their ability to cope with dynamic environments. DRs cope well with dynamic environments as decisions are made as and when a machine becomes available to process a job's operation. When a machine finishes processing a job, each job currently waiting in the machine's queue is evaluated and assigned a priority value by the DR. The job with the highest priority is selected to begin processing next. One disadvantage of DRs is that they take into account the local and current conditions of the shop, without considering the future implications of decisions made. Another disadvantage of DRs is it is a time consuming (and very hard) process to construct them manually.

Genetic Programming (GP) [14] is an evolutionary computation technique which has been shown to be sufficient in representation and search capability to automatically discover DRs successfully [9,11,16]. The mathematical representation of dispatching rules maps naturally to tree-based GP [14]. Each GP individual represents a DR, and through evolution the fitness of the individual is assessed through discrete-event simulations of job shops in which the individual is used to dispatch jobs. There has been work to encourage the evolution of "less-myopic" DRs through the inclusion of properties from the wider shop [10], which showed that the inclusion of additional "less-myopic" terminals improved the performance of DRs on scenarios with high utilisation, and reduced the expected queue length at machines. This paper investigates "less-myopic" in a different way, by including an assessment of how well a DR makes local decisions over a longer decision horizon in its fitness evaluation.

Local search algorithms try to improve on an initial solution by searching over a defined neighbourhood of the initial solution [1]. In static JSS environments, a DR can be used to create a schedule for processing the jobs, and local search can be applied to improve the schedule. This is *not* how we are using local search in this paper. The main goal of this work is to modify a GP based system for automatic generation of dispatching rules to incorporate a local search element into the fitness evaluation stage of the GP system to provide more feedback to DRs on their scheduling performance, to encourage a more global perspective in the evolved rules. In this paper we use local search only to evaluate the fitness of DRs, *not* to change the order in which jobs are dispatched.

When a machine becomes available, a DR builds an initial schedule of jobs at the current machine. This is evaluated over an *extended decision horizon* of the current machine and the next machine of each job. Local search is used to attempt to improve that sequence, and information on the possible improvement contributes to the fitness of the DR.

We want to find DRs that make good decisions, which we can do by looking at the wider local effect when dispatch decisions are made, and investigate whether this can encourage the evolution of DRs which are less-myopic, and *better* at scheduling jobs in an order which is *better* beyond just the first job in the queue, with *better* generalisation performance on unseen problem instances.

The remainder of this paper is organised as follows. Section 2 describes background relating to JSS and GP. Section 3 describes how local search is incorporated into our GP method for automatic discovery of DRs. Section 4 describes the experimental design and Sect. 5 discusses the results. Section 6 concludes the paper and provides directions for future work.

2   Background

Dynamic Job Shop Scheduling Problem. In this paper we are interested in the dynamic job shop. Jobs arrive according to a stochastic process, and we have no knowledge about the job until it has entered the shop system. Each job j arrives at time r_j with a sequence of N_j operations, $O_j = (\sigma_{j,1}, \ldots, \sigma_{j,N_j})$, in a predefined route, and where $1 \leq N_j \leq n$. The ith operation of job j, $\sigma_{j,i}$, has processing time $p(\sigma_{j,i})$ on machine $m(\sigma_{j,i})$. Each job joins the queue at the first machine on its route, $m(\sigma_{j,1})$, upon its arrival into the shop. Each job has a due date d_j and an importance weighting w_j. The eventual completion time of a job is denoted C_j and job tardiness is then $T_j = \max\{0, C_j - d_j\}$.

A DR is used to select which of the jobs currently waiting in the queue at a particular machine will be processed next using various properties of the job, machine and shop. Once an operation σ has completed its processing time on a machine it is moved to the next machine on its route, or if the entire job is completed then it is delivered to the customer. There are many different objectives that can be optimised in JSS; we are interested in minimising the total weighted tardiness:

$$TWT = \sum_j w_j T_j = \sum_j w_j \max\{0, C_j - d_j\}. \tag{1}$$

Existing DRs. There are many DRs in the literature. No DR is known to outperform others across all objective functions and shop environments. Some of the simplest rules are shortest processing time (SPT) which schedules the job with the shortest processing time next, and its weighted counterpart, weighted shortest processing time (WSPT), which schedules the job with largest $w/p(\sigma)$. These are components of the Apparent Tardiness Cost (ATC) [21] and the weighted version of COVERT (Cost over Time) [21] dispatching rules, which have good performance for dynamic job shops with the TWT objective. ATC assigns job j the priority

$$\frac{w_j}{p(\sigma_{ji})} \exp\left(-\left[\frac{d_j - t - p(\sigma_{ji}) - \sum_{q=i+1}^{N_j}((b \times p(\sigma_{jq})) + p(\sigma_{jq}))}{kp_{avg}}\right]^+\right)$$

where t is the current time and p_{avg} is the average processing time of the waiting jobs at the machine. Leadtime estimation parameter b is fixed at 2.0 and lookahead parameter k is set to 3.0 as in [21]. The wCOVERT rule assigns priority values as the expected tardiness cost per unit of imminent processing time:

$$\frac{w_j}{p(\sigma_{ji})} \left[1 - \frac{(d_j - t - \sum_{q=i}^{N_j} p(\sigma_{jq}))^+}{k \sum_{q=1}^{N_j}(b \times p(\sigma_{jq}))} \right]^+$$

where the lookahead parameter $k = 2$. The $[A]^+$ notation takes the maximum of A and 0, this means wCOVERT assigns priority 0 if the slack "exceeds some generous 'worst case' estimate of the waiting time" [21].

GP for Automatic Generation of Heuristic Dispatching Rules. Heuristics are methods that seek to find good quality solutions in reasonable computational time; optimality cannot be guaranteed [4]. Dispatching rules can very naturally be represented as trees, as in tree-based GP. Recent research has focused on developing new DRs using approaches such as genetic programming [4]. GP searches the space of heuristics rather than searching the solution space directly [4], as the GP trees do not represent schedules, but heuristic DRs.

GP has been used to develop DRs in a range of scheduling environments, from the simpler static two-machine flow shop [6] and job shop [9], to complex dynamic environments, e.g., in semi-conductor manufacturing [18], and varying objectives such as makespan [11,12], mean flow time [8], and total weighted tardiness [12]. Rules evolved by GP are frequently reported to compete with rules from the literature [11,16].

Local Search. Local search is one heuristic approach often used on static scheduling problems. Local search provides a "robust approach to obtain high-quality solutions to problems of a realistic size in a reasonable time" [1]. Local search methods start with an initial solution and try to find better solutions by searching neighbourhoods. Some basic neighbourhood definitions for local search that can be applied to schedules are transposition, insertion and swap [1]. Let n be the number of jobs in the queue of the machine. *Transpose* swaps two adjacent jobs in the queue (neighbourhood has size $O(n)$). *Insert* moves a job from one position to another and *Swap* swaps two jobs that can be anywhere in the queue (both have neighbourhood size $O(n^2)$). With dynamic scheduling problems, the jobs available to be scheduled are frequently changing with the arrival of jobs from outside the shop system and from other machines. This makes the application of local search difficult, as we do not know when another job will arrive, or any of its properties, therefore an optimised queue order is unlikely to still be optimal when more jobs arrive, or when we consider the final objective value function.

Instead of applying local search to find better DRs with a different order of jobs, this paper will use local search in the fitness evaluation process to improve the performance of the DRs. Details will be discussed in the next section.

Table 1. Terminal set used in GP system.

Feature	Symbol	Feature	Symbol
Job Properties		*Machine Properties*	
Processing time of operation	PR	Ready time of machine	RM
Remaining processing time of job	RT	Number of jobs in queue	NQ
Remaining number of operations	RO	Average wait time of last	QW
Ready time of job	RJ	Five jobs processed	
Due date of job	DD	Number of jobs in queue	NNQ
Weight of job	W	At the next machine job visits	
Next operation's processing time	NPR	Average wait time of last five jobs processed at next machine the job visits	NQW
Shop Properties			
Current time	CT	Average wait time of last five jobs processed across all machines	AQW

3   The New Method

Here we describe our new method, which uses local search as an additional evaluation of the fitness of a given DR, across an extended decision horizon. We use local search to evaluate potential queue orders, and compare these to the priority sorted queue. We are interested to see if the increase in computational time is a reasonable trade off for better evolved DRs.

3.1   GP System

GP individuals are DRs. The function set is $\{+, -, \times, \%, \texttt{if>0}, \texttt{max}, \texttt{min}\}$. All arithmetic operators take two arguments. $+$, $-$ and $\times$, have their usual meanings and the $\%$ operator is protected division, returning one if dividing by zero. The $\texttt{if>0}$ function takes three arguments; if the first argument is greater than 0 then it returns the second, else the third is returned. The max and min functions take two arguments and return the maximum and minimum of their arguments respectively. The properties of jobs, machines and the job shop that are used as terminals of the GP system are given in Table 1. The properties, NPR, NNQ and NQW all return zero if the job's current operation is its last. If fewer than five jobs have visited a machine, then WQ, NQW and AQW return the average wait time of the jobs which have visited, and if the machine has not yet processed any jobs then 0 is returned [10].

Fitness. The fitness of a DR (individual) in the current GP population is evaluated using discrete-event simulations of problem instances of a job shop. In each method the fitness value is calculated on four training problem instances, and the mean of the fitness values across these problem instances used is the fitness of the DR. In the benchmark GP method, the objective of interest to be minimised, TWT, is normalised by the expected utilisation and this is used as the fitness value. Each of our new methods use a different fitness value.

3.2   Local Search for Evaluating Dispatching Rules

We consider four different ways of using local search as a contributor to the fitness of an individual. We take the priority sorted queue of jobs waiting at the time the machine needs to make the scheduling decision as the initial solution. The neighbourhood search operator used is *SwapFront*, which swaps the job at the front job with each other job. This is used as it is a simple operator with small neighbourhood size, $(n-1)$ for n jobs in the queue of the machine, and therefore low computational cost. The evaluation process is shown in Fig. 1.

We compare neighbourhoods by calculating the expected contribution to total weighted tardiness (TWT) of the queued jobs, which is the sum of weighted expected tardiness of each of the jobs in the queue. We calculate this by taking the expected completion time of each job given its current position in the queue at the current machine, and where it would fit in the queue of jobs at the next machine the job visits (assuming the current state of the job shop as static, with no new arrivals except those moving from the current queue). For each job j in the queue, this is calculated as

$$E(C_j) = \mathrm{CT} + QC_j + \mathrm{PR}_j + QN_j + \mathrm{NPR}_j + QR_j + (\mathrm{RT}_j - \mathrm{PR}_j - \mathrm{NPR}_j) \qquad (2)$$

For a given job d_j, CT, PR, NPR, RT are constants as in Table 1. QC_j is the time remaining waiting (queuing) at the current machine (M_C), i.e., the sum of the processing times of jobs ahead of job j in the queue under the current ordering of jobs. QN_j is the time spent waiting (queuing) at the next machine on the job's route, (M_N). This is determined by treating the next machine's queue as a one machine problem with arrivals only from the current machine and using the DR to dispatch jobs until all jobs that join this machine from the current machine have been dispatched. This gives us a lower bound on the length of time each job from machine M_C that next visits M_N is expected to be in the queue at M_N, as in the full simulation jobs may arrive into the shop and from the other machines. QR_j is the sum of the average expected waiting times at each remaining machine on the job's route after machines C and N. If a job does not have operations remaining after the current or next machine then QN_j and QR_j will have value 0. Further we are only interested in changing the order of jobs at the first machine, and calculating the predicted time in the next queue, therefore for each job QR_j is also a constant. As our objective is to minimize the total weighted tardiness, we are seeking to minimize Eq. (3) across the neighbourhood being searched.

$$Total = \sum_{j=1}^{J} w_j \times \max\{0, E(C_j) - d_j\}. \qquad (3)$$

Each time the DR is applied to select the next job, the expected contribution of the original queue order, $Total_0$, is calculated. If $Total_0 = 0$ then we cannot improve on the current queue ordering, so we do not apply local search. For each neighbouring solution, $Total$ is calculated, so we can find the minimum of all neighbourhoods searched, $Total_{min}$, which is used in the next stage.

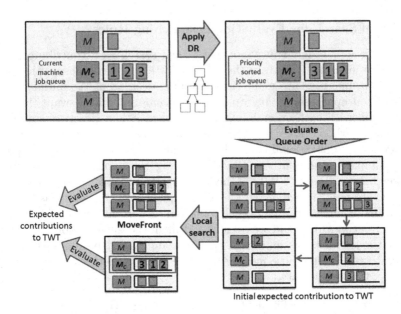

Fig. 1. Diagram of evaluation process when a machine becomes available to process a job.

Local Search to investigate Local Decision Making. When we apply local search and the new job queue order has an improved (smaller) expected contribution to TWT than the original, we calculate the difference $penalty = Total_0 - Total_{min}$. We sum all the penalties incurred during the discrete event simulation, and average over the number of times the local search was applied; this gives $penalty_{mean}$.

- *Local Search Single Objective (LS-SO).* We use local search to evaluate the job queue every twentieth time the DR is called to select a job for dispatch. The overall fitness of a DR for a problem instance is the sum of TWT and the penalty, $TWT + penalty_{mean}$.
- *Local Search Multi Objective(LS-MO).* Local search evaluates the job queue every twentieth time the DR is called to select a job for dispatch. This is a multi-objective method based on LS-SO, using the NSGA-II algorithm [5]. The penalty is used as a distinct second objective. The first objective is TWT and the second objective is $penalty_{mean}$.

Local Search to investigate Tie Breaking (TB). These methods investigate how often a DR assigns the same priority value to different jobs, and whether the default tie break, of using the SPT rule, is selecting the best job in those situations. Local search is used among the sub-schedule of jobs which have been assigned the same highest priority value in the queue. Sub-schedules are compared based on their expected contribution to the objective function. We run two discrete-event simulations for each problem instance. In the first we do not use local search, and ties are broken using SPT, this gives TWT_0. In the second we use local search to alter which of the jobs that are tied for top priority is dispatched, dispatching the job which leads to the lowest expected contribution to TWT, and the resulting final value of the objective function is TWT_{LS}. Due to computational restraints we use local search only every second time a tie occurs.

- *Tie Breaking Single Objective (TB-SO)*. This variant uses single objective GP, combining the TWT fitness with an added penalty if the performance of the DR was better with local search, i.e., if $TWT_{LS} < TWT_0$. The fitness of an individual for a problem instance is $TWT_0 + \max\{TWT_0 - TWT_{LS}, 0\}$.
- *Tie Breaking Multi Objective (TB-MO)*. This variant uses a multi-objective approach. The first fitness objective is TWT_0, the second fitness objective is the penalty for possible improvement on the schedule the DR produced, $\max\{TWT_0 - TWT_{LS}, 0\}$. We use the NSGA-II algorithm [5].

We want to find out if TB-SO and TB-MO can reduce the number of ties in test cases, and if the TWT performance is improved.

4   Experiment Design

We implement a GP system using ECJ20 [15] for evolving DRs, which are represented by the GP trees.

Problem Instances. We randomly create problem instances of job shops with ten machines. We present results on one test and one extreme test scenario. The extreme testing to see how well the rules generalise to a job shop with different distribution and priority assignment. We are dealing with dynamic JSS, so jobs arrive stochastically according to a Poisson process with rate λ for all problem instances. The settings for these are shown in Table 2. The desired expected utilisation of machines (ρ) is achieved by setting $\lambda = \dfrac{\rho}{(\mu \times p_M)}$, where p_M is the expected number of machines a job will visit (i.e. the expected number of operations in each job). Job due dates are set by [2], $d_j = r_j + h \times \sum_{l=1}^{N_j} p(\sigma_{j,l})$, where h is a due date tightness parameter, randomly chosen with equal probability from the choices available for each job. Jobs are given weight 1, 2 or 4, with probability $(0.2, 0.6, 0.2)$ [16]. Four training scenarios are used. The processing times at each machine follow a discrete uniform distribution with mean μ, i.e., $U(1, 2\mu - 1)$. A warm up period of 100 jobs is used, and we collect data from the next 200 jobs to arrive ($N = 200$), however new jobs keep arriving in the system until the 300th job is completed. This is a very low number of jobs due to the increase in computational time required by local search. In testing, we increase the number of jobs we collect data to $N = 1000$. In testing scenario T1, all jobs have due date tightness of 4, and utilisation is 0.95. In extreme testing, XT1, the processing times follow a geometric distribution with mean $\mu = 25$ (parameter $p = 0.04$), and utilisation is 0.95. We also change the weights given to jobs, including an additional weight for *very* important jobs; jobs are now given weight 1, 2, 4 or 8, with probability $(0.2, 0.5, 0.2, 0.1)$. Due date tightness is equally likely from $\{2, 2.5, 3\}$; these are the same or tighter than in training.

Table 2. Simulation properties for training and testing.

		μ	ρ	h	Operations			μ	ρ	h	Operations
Training	TR1	25	0.90	$\{2,3,4\}$	Full	Training	TR3	25	0.90	$\{2,3,4\}$	Missing
	TR2	25	0.95	$\{2,3,4\}$	Full		TR4	25	0.95	$\{2,3,4\}$	Missing
Testing	T1	25	0.95	4	Full	Extreme testing	XT1	25	0.95	$\{2,2.5,3\}$	Full

GP System. The initial population is generated using the ramped-half-and-half method [14]. The population size is 50, to restrain the computational time due to the incorporation of local search, and evolution is for 50 generations, a standard setting. GP trees have a maximum depth of six. Genetic operators crossover, mutation and elitism, use rates of 85 %, 10 % and 5 % respectively. Tournament selection with a tournament size of seven is used to select individuals for genetic operators. This is a common setting that has been previously used [17].

5   Experimental Results

We performed 50 independent GP runs for each method, with the same seed. From these the best-of-run individual (LS-SO or TB-SO), or final non-dominated front (LS-MO or TB-MO) are tested on test and extreme test instances.

5.1   LS-SO and LS-MO

Figures 2 and 3 present the test results of the evolved DRs on T1 and XT1. These methods are compared to a benchmark method which only used the local search penalty calculation during testing. In particular the zoom-in on the overall best front of DRs in Fig. 2 shows that the LS-MO method was able to find a large number of DRs which attain lower TWT on both test instances. Four DRs form the front on this scenario. The lowest TWT attained by LS-MO and LS-SO methods are lower than the lowest attained by the benchmark.

Including local search in the evaluation process has found DRs which attain good performance. There are a large number of DRs from LS-MO and LS-SO that achieve lower penalty than the benchmark for similar TWT values, particularly for the lower TWT values. This shows that including local search has improved the local performance of DRs.

5.2   TB-SO and TB-MO

Figures 4 and 5 present the performance of TB-SO and TB-MO on two test scenarios. These methods are compared to a benchmark method which only used the tie breaking penalty calculation during testing. The TB-SO method produced the DRs which attained the lowest TWT value on both scenarios. Most of the best evolved DRs have a penalty of 0, this supports our hypothesis that classifiers which are better at separating jobs by assigning distinct priorities

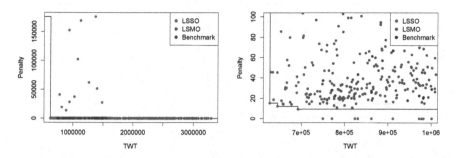

Fig. 2. Graph of the combined non-dominated front from non-dominated fronts from LS-MO and best-of-run individuals from LS-SO on scenario T1 with close up of lowest TWT attained on right.

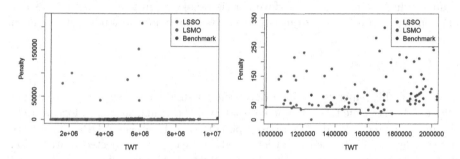

Fig. 3. Graph of the combined non-dominated front from non-dominated fronts from LS-MO and best-of-run individuals from LS-SO on scenario XT1 with close up of lowest TWT attained on right.

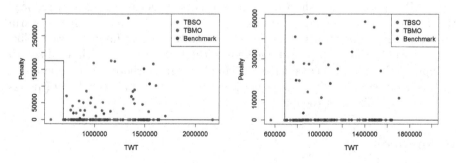

Fig. 4. Graph of the combined non-dominated front from non-dominated fronts from TB-MO and best-of-run individuals from TB-SO on scenario T1 with close up of lowest TWT attained on right.

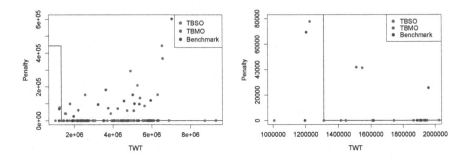

Fig. 5. Graph of the combined non-dominated front from non-dominated fronts from TB-MO and best-of-run individuals from TB-SO on scenario XT1 with close up of lowest TWT attained on right.

are more effective. The spread of TWT and penalty results on both scenarios is similar, which suggests that performing tie-breaking with local search does not offer enough improvement to justify the additional computational cost incurred.

6   Conclusions

The goal of this paper was to investigate the possible improvements to training of DRs for the dynamic ten-machine job shop in GP through the use of local search. We implemented a local search based penalty, which punished DRs which were not scheduling jobs in the best order based on the current projected contribution to the TWT objective function. Results show that the inclusion of the local search penalty in evaluation has led to the evolution of DRs which have better local performance, and achieve some better TWT values. This is worth further investigation. The inclusion of local search for tie-breaking did not offer enough improvement in TWT in its current implementation to warrant its use.

Due to the computational cost we have investigated small examples, with a small GP population and short problem instances. This has significantly reduced the number of scheduling decisions made in training, and with the shorter warm up period the system may not have reached steady state. In future we would like to increase the warm up period and reduce the frequency in which local search is applied. We will also consider other simple neighbourhood search operators, such as *Transpose* (swaps two adjacent jobs in the queue) and *MoveFront* (inserts each job at the front of the queue).

References

1. Aarts, E., Lenstra, J. (eds.): Local Search in Combinatorial Optimization. Wiley, Chichester (1997)
2. Baker, K.R.: Sequencing rules and due-date assignments in a job shop. Manage. Sci. **30**(9), 1093–1104 (1984)

3. Blazewicz, J., Domschke, W., Pesch, E.: The job shop scheduling problem: conventional and new solution techniques. Eur. J. Oper. Res. **93**(1), 133 (1996)
4. Burke, E.K., Hyde, M.R., Kendall, G., Ochoa, G., Ozcan, E., Woodward, J.R.: Exploring hyper-heuristic methodologies with genetic programming. In: Mumford, C., Jain, L. (eds.) Computational Intelligence, Intelligent Systems Reference Library, vol. 1, pp. 177–201. Springer, Heidelberg (2009)
5. Deb, K., Pratap, A., Agarwal, S., Meyarivan, T.: A fast and elitist multiobjective genetic algorithm: NSGA-II. IEEE Trans. Evol. Comput. **6**(2), 182–197 (2002)
6. Geiger, C., Uzsoy, R., Aytug, H.: Rapid modeling and discovery of priority dispatching rules: an autonomous learning approach. J. Sched. **9**(1), 734 (2006)
7. Hart, E., Ross, P., Corne, D.: Evolutionary scheduling: a review. Genet. Program. Evolvable Mach. **6**, 191–220 (2005)
8. Hildebrandt, T., Heger, J., Scholz-Reiter, B.: Towards improved dispatching rules for complex shop floor scenarios: a genetic programming approach. In: Proceedings of the 12th Annual Conference on Genetic and Evolutionary Computation, pp. 257–264 (2010)
9. Hunt, R., Johnston, M., Zhang, M.: Evolving machine-specific dispatching rules for a two-machine job shop using genetic programming. In: Proceedings of the IEEE Congress on Evolutionary Computation, pp. 618–625 (2014)
10. Hunt, R., Johnston, M., Zhang, M.: Evolving less-myopic scheduling rules for dynamic job shop scheduling with genetic programming. In: Proceedings of the Genetic and Evolutionary Computation Conference, pp. 927–934 (2014)
11. Jakobović, D., Jelenković, L., Budin, L.: Genetic programming heuristics for multiple machine scheduling. In: Ebner, M., O'Neill, M., Ekárt, A., Vanneschi, L., Esparcia-Alcázar, A.I. (eds.) EuroGP 2007. LNCS, vol. 4445, pp. 321–330. Springer, Heidelberg (2007)
12. Jakobovi, D., Marasovi, K.: Evolving priority scheduling heuristics with genetic programming. Appl. Soft Comput. **12**(9), 2781–2789 (2012)
13. Jones, A., Rabelo, L.C.: Survey of Job Shop Scheduling Techniques. Technical report, National Institute of Standards and Technology, Gaithersberg (1998)
14. Koza, J.R.: Genetic Programming: On the Programming of Computers by Means of Natural Selection. MIT Press, Cambridge (1992)
15. Luke, S.: Essentials of Metaheuristics, 2nd edn. Lulu (2013). http://cs.gmu.edu/~sean/book/metaheuristics/
16. Nguyen, S., Zhang, M., Johnston, M., Tan, K.C.: A coevolution genetic programming method to evolve scheduling policies for dynamic multi-objective job shop scheduling problems. In: Proceedings of the IEEE Congress on Evolutionary Computation, pp. 3261–3268 (2012)
17. Nguyen, S., Zhang, M., Johnston, M., Tan, K.C.: A computational study of representations in genetic programming to evolve dispatching rules for the job shop scheduling problem. IEEE Trans. Evol. Comput. **17**(5), 621–639 (2013)
18. Pickardt, C.W., Hildebrandt, T., Branke, J., Heger, J., Scholz-Reiter, B.: Evolutionary generation of dispatching rule sets for complex dynamic scheduling problems. Int. J. Prod. Econ. **145**(1), 67–77 (2013)
19. Pinedo, M.L.: Scheduling: Theory, Algorithms, and Systems, 3rd edn. Springer, Heidelberg (2008)
20. Potts, C., Strusevich, V.: Fifty years of scheduling: a survey of milestones. J. Oper. Res. Soc. **60**(S1), 41–68 (2009)
21. Vepsalainen, A., Morton, T.: Priority rules for job shops with weighted tardiness costs. Manage. Sci. **33**, 1035–1047 (1987)

Author Index

Printed in the United States
By Bookmasters